What You Need To Know

about your

GOLD & SILVER

ISBN 0-918080-44-4 (softcover edition)

Distributed By
Gem Guides Book Co.
3677 San Gabriel Parkway
Pico Rivera, Calif. 90660

Third Printing

Photography
by
Oscar T. Branson

Sketches
by
Connie Asch

TREASURE CHEST PUBLICATIONS, INC.
P.O. Box 5250 — Tucson, Arizona 85703

Contents

Foreword

The purpose of this book is to endeavor to inform the reader how to recognize gold and silver and their alloys, how to make simple tests to identify them and how to determine their quality and value. To make these tests it will be necessary to use only two acids and two chemicals which, if handled carefully, will cause no problems. If handled carelessly, they could result in pain and injury, and damage to almost anything with which they come in contact. These acids and chemicals are poisonous and are deadly if ingested, even in small amounts. Without using these acids and chemicals, there is no way to determine the genuineness or quality of gold or silver. Jewelers, gold and silver workers and chemists have been using these chemicals continuously for many decades, with very few accidents or casualties. There will be repeated warnings throughout the book when the use of these chemicals is indicated to perform tests or determinations.

The author and publisher take no responsibility whatsoever if or when the use of these chemicals results in damage to persons or property. The person buying or using these chemicals assumes the full responsibility for damage to himself and others regardless of the consequences.

To identify and evaluate the precious metal in a few pieces of jewelry it is not necessary to buy and use the acids and equipment described in this book. Most jewelers and precious metal buyers have these acids and equipment. They will be glad to show how they test and explain how to evaluate the articles of gold and silver you may offer for sale or appraisal.

Simple easy tests will identify your gold and silver.

No knowledge of chemistry, science or metallurgy is necessary to understand the explanations and make the tests described in this book. To perform these simple tests requires no more mental or physical dexterity than it takes to fry an egg.

Gold

Gold is the most beautiful of all metals and probably the earliest metal discovered by man. It is the only truly yellow metal found free and almost pure in nature. It is non-magnetic and one of more than one hundred elements that make up the planet Earth. Gold is the most important of the precious metals and is, and always has been, of the greatest economic importance. Many nations base the value of their currency on the amount of gold in their treasuries. Gold is the only universally accepted medium of exchange. Pure gold is soft, only about 2.5 to 3 in hardness (mohs scale), almost too soft to be used in jewelry without being alloyed with another metal. It is harder than lead (1½). Lead can be scratched with the fingernail, gold cannot. It is the most ductile and malleable of all metals and can be beaten into a sheet or leaf so thin that one ounce will cover an area of 160 square feet. It is a very heavy metal, the specific gravity is 19.3, only slightly lighter than platinum (21.4 s.g.) and almost twice as heavy as silver (10.5 s.g.).

It is this great weight that makes the easy recovery of gold possible. When gold and the rocks and sand with which it is found in nature is shaken or agitated in a pan with water, the gold quickly settles to the bottom of the pan. Many millions of ounces of gold have been recovered from stream sands and gravels by this method. A cubic foot of gold weighs 1,204.8 pounds. A cubic inch weighs about 10 ounces avoirdupois. Gold is almost completely insoluble in nature and in ordinary acids. Therefore it is tasteless, making it ideal for use in dental work. It is soluble only in a mixture of hydrochloric and nitric acids, called "aqua regia", and in some cyanide solutions.

These two characteristics of gold, being insoluble in and unaffected by nitric acid, and soluble in aqua regia, are what make gold easy to test for and easy to recover and purify.

The Price Of Gold

The price of gold is set daily at the gold market in London by members of a number of large institutions dealing in gold. The buyers and sellers actually determine the price which depends upon supply and demand, and the different amounts that are paid for gold cause the price fluctuations during the day. If there are more orders to purchase gold than there are contracts to sell, the price goes up. If there are more contracts to sell than there are orders to buy, the price offered is usually lower. The published price is what gold was selling for at the end of the trading day in London and is in U.S. dollars per troy ounce. The daily price quote can be obtained from banks, commodity brokers, coin and precious metal dealers and the financial page of the local newspaper.

The daily price quote is based upon the sale of the large 400 ounce (27.4 lbs. av.) sometimes called the "Good Delivery" bar. This is the universally recognized medium of exchange between large banks and nations. It is the largest bar available of almost pure gold (.995). With gold at $300 per ounce, it's value would be about $120,000. It is the only way gold can be bought at near market price. There is usually little or no premium charged in the sale of these "goodies", only the shipping and handling charges. Any amount smaller than this bar will have premiums, commissions or brokerage fees added to the price. These large bars are usually kept in special bank vaults until needed in industry or, if traded or purchased by governments, are merely moved from one vault to another, sometimes only a few feet away.

The next smaller size bar is the 100 ounce (6.85 lbs. av.) bar and, if the price of gold is $300 an ounce, the base price of this bar will be about $30,000. These bars are offered in several grades of fineness (995, 999+, 999.5 and 999.9). The higher the quality, the higher the price per ounce. Gold of any quality higher than .995 is electrolytically refined and will sell at a much higher price per ounce. There will also be a 2 or 3 percent premium in addition to the handling and delivery charges for these bars.

The most popular size bar and the easiest to transport is the one kilo bar 32.15 ounces (2.2 lbs. av.). The base price is about $9,645 if gold is $300 an ounce. These bars are also produced in several grades of fineness, .995 and above. There will always be a premium or mark up over the gold price on these bars, in addition to the handling and shipping charges.

Almost all gold bars produced by reliable refineries are stamped with the fineness, weight and registration or serial number.

There are also a number of bars, from one ounce to 10 ounces in weight, produced by refiners. In addition to being stamped with the fineness, weight and registration number, they usually have the name of the refiner or bank which is issuing them. These bars are produced for the small investor for storing in the safety deposit box or safe at home and are considered by many to be a sure hedge against inflation or devaluation of paper money.

The premium charged is high, usually 5 to 10 dollars an ounce over the market price of gold. When these small bars are offered for resale, the premium is lost and sometimes a commission is charged for the sale by the bank or broker.

If the large gold bars have been removed from the bank storage vault and later sold, it is very possible a buyer might insist on an assay to prove the value or fineness of the bar.

There are a multitude of smaller gold bars or wafers being offered for sale by banks, jewelers, coin shops and department stores. These are offered in sizes from ¼ ounce to one ounce. The sizes of less than ½ ounce are usually marked 24 karats and not the percentage of fineness. This places them in the class of jewelry and if they have a bale or ring soldered on they can be, by law, 1 karat less than they are stamped. The price you can expect to pay for smaller ones is about what you would pay for gold jewelry, sometimes 2½ times the value of the gold they contain.

Selling Gold And Silver

On December 31, 1974, all gold regulations in the United States were abolished. We can now buy, sell, keep, melt or do anything we wish with as much gold as we please, or can afford.

The United States government (the U.S. Treasury Department or the U.S. Mint) is not buying gold or silver and has not done so since March of 1968.

Gold and silver in small amounts can be sold directly for immediate payment to a number of different businesses such as jewelry stores, jewelry supply houses, coin dealers, precious metal buyers and some small refiners. Usually large precious metal refiners are not interested in buying in small quantities. You need to have 100 ounces of gold bearing material or 500 ounces of silver bearing material for it to be profitably refined.

The best price will be obtained from the more reputable businesses. Caution should be used when selling to transitory one-day-only or short time motel room operators. Shop around, any legitimate buyer will not hesitate to give you an offer and explain his calculations. If you are not satisfied with his offer, go to someone else.

Due to market price fluctuations, offers may vary from day to day. You can obtain the market price from the financial page of your daily newspaper or by phone from a commodities broker.

Banks seldom buy gold or silver in any form, although they might possibly act as intermediaries for buying or selling bullion or trade-gold coins, such as the Krugerand. Of course they charge a commission for either transaction.

There is usually quite a difference in the theoretical value of the calculated amount of gold in an object and the actual amount that will be paid for it by a gold or silver buyer. There are several reasons for this difference:

The actual gold in an article stamped with a certain karat mark is almost always about one karat less than it is marked.

The gold and silver buyer is a middleman and must make a profit for the use of his time, knowledge and investment. The gold buyer will have to accumulate a large amount of gold scrap, sometimes thousands of dollars worth, before it is profitable to ship it to the refiner. The refiner cannot accept small amounts of gold to refine because of the cost of the labor, time required and the high cost of the chemicals required. The cost of refining a small amount of gold would probably be as much as the value of the gold itself.

The larger the amount sent to the refiner, the lower the refining cost for each ounce refined. The most economical amount would be 50 ounces or more. The refining charge can be as much as 10 to 20 percent of the value of the refined precious metal.

The price of gold varies from day to day and the scrap buyer usually pays a price based on the daily quote. This means he takes a risk because, if the price of gold drops drastically before he can have it refined, he will take a loss. Also a risk is taken in transit when the metal is shipped to the refiner. When everything is considered, the gold scrap buyer should be entitled to a good profit. Most buyers require a profit of 20 to 30 percent.

The refiner does not pay the metal buyer for the precious metals in the shipment until after it is refined and the value of the metals which were contained in the shipment can be calculated. It also takes time to refine a shipment of precious metal scrap, usually six weeks, and if the refiner is extremely busy, it could be several months before payment will be made. This means that the person selling the scrap gold or silver to a metal buyer might possibly receive only 50 to 60 percent of the value of the precious metal in the article.

Gold Jewelry As An Investment

Is gold jewelry a good investment? This depends on whether you consider it just as an investment in gold or also as an investment in the pleasure you receive from wearing beautiful gold jewelry. Most gold jewelry is only 10 to 14 karat and the best is 18 karat. Jewelry made of pure gold would be too soft for wearing. The articles of jewelry that you will buy have a mark up of 100 to 250 percent over the actual value of the gold content. For example, if you bought a 14 karat man's ring that weighs one half ounce, it would contain about one third of an ounce of pure gold. If the market price of pure gold was $300 an ounce, the gold in the ring would be worth about $87 and the ring would probably cost about $200. Much of this cost is in the labor to make the ring and the profit for the jeweler who sold it. If you would try to sell the ring after wearing it for some time, you might only get the value of the pure gold that it contains. So gold jewelry cannot be considered an investment for the gold alone, except, of course, when there is a dramatic increase in the price of gold. Then it becomes a good investment. Also, if the joy and pleasure of wearing beautiful gold jewelry is considered as important as the value, then it should certainly be considered a good investment.

The best method for testing gold and silver is with nitric acid. The easiest and safest way is to use a small glass bottle with a glass stopper. These bottles have a small glass stopper ground to fit the neck of the bottle to prevent leakage of the acid and are called acid testing bottles. The bottom of the stopper is elongated into a glass rod, so that only a drop of acid can be taken from the bottle at a time and applied to an object suspected of being gold or silver. The bottle should be one half ounce capacity and it is safest to have only about one fourth ounce or less of acid in the bottle. Nitric acid is a very poisonous, dangerous and corrosive liquid and should be handled with extreme care at all times. When it is allowed to touch the skin, a sharp burning pain is experienced and a burn results. The contaminated area turns bright yellow and within a few days that area of skin peels away. If the skin is touched with acid, flush or dip instantly in water. This should be done very quickly to prevent an acid burn. When working with acid it is advisable to wear glasses or safety glasses to protect the eyes.

To determine if a gold colored article is really gold, first clean the surface to be tested of any coating or transparent varnish, plastic or lacquer. Some new articles, or articles that have not been worn, may still retain such a surface. Acid usually takes a long time to react through this type of coating. This coating may be removed by using a solvent such as lacquer thinner, acetone or fingernail polish remover or by rubbing with fine emery cloth. When you are sure the true surface is exposed, put a drop of nitric acid on the area to be tested. The higher karat golds 14, 16, 18, 20, 22, and 24 karat will show no change. Gold of 12 karat will show a very slight change, but 10 karat or lower will change to a brown color where a drop of nitric acid has been applied.

To become acquainted with the reaction of nitric acid on articles of different karat quality, it will be necessary to try the test on articles of known or marked karat value, especially some examples of 10 and 12 karat so you can see the color change caused by the acid. These same articles can also be used when learning to tell the karat of the gold by using the touch stone.

It is reasonable to assume that articles stamped with a karat mark are really of that quality, but since there are many tricksters in the trade, it is wise to test all articles. If the quality of gold is not determined correctly, a mismarked article could result in a substantial loss, especially with the high price of gold. If the metal is yellow and there is no color change with nitric acid, you can be reasonably sure it is a gold alloy. There is no other yellow metal that shows no reaction to nitric acid. If the metal is white and it is stamped with a karat value, you can be fairly certain that it is white gold. If it is a white metal and is not stamped with a karat mark, there is a good chance it is one of a large number of alloys other than gold which do not show a change with nitric acid. Some of these white alloys are stainless steel, platinum, palladium, certain aluminum alloys and many others. The presence of gold in a white metal alloy can easily be determined and will be explained later in this book. Silver alloys are also white but DO show a change when a drop of nitric acid is applied to the surface. When it is determined that the surface of the article in question is gold, it will be necessary to find out if the article is solid or gold plated. This is necessary because many articles were plated and stamped with the karat mark of the plate or coating but not stamped "Plate" or "Plated". For many years the laws, if any, governing the manufacturing of gold articles were lax, hard to control and few or no penalties were imposed for violations. Therefore accept only what checks out to your satisfaction.

The Touchstone Method For Testing Gold

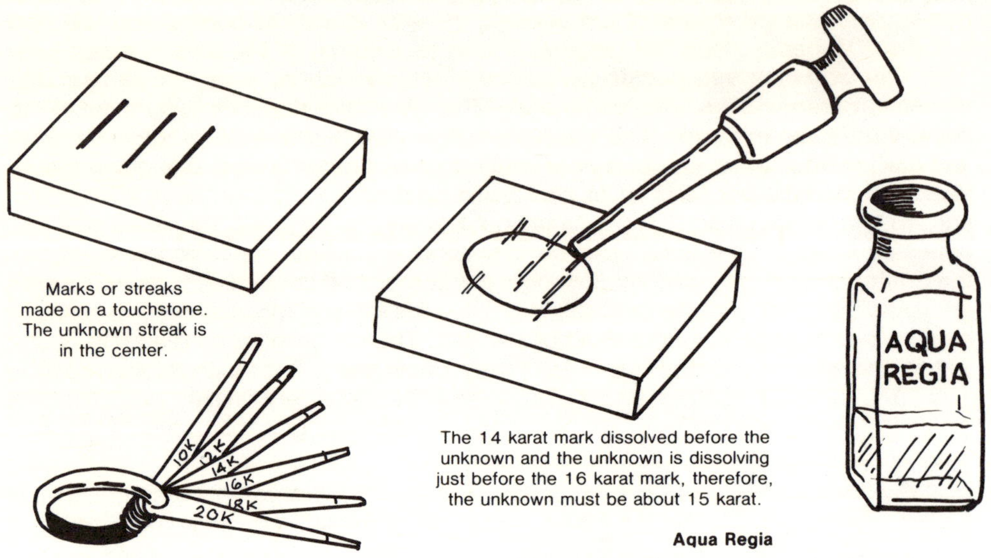

Marks or streaks made on a touchstone. The unknown streak is in the center.

The 14 karat mark dissolved before the unknown and the unknown is dissolving just before the 16 karat mark, therefore, the unknown must be about 15 karat.

The standard yellow gold set of testing needles. These small brass wedges are marked with the karat of the gold on the tips.

Aqua Regia

With a medicine dropper measure out 8 drops of water into a one half ounce acid testing bottle. Add two drops of hydrochloric acid; then using a clean dropper add 10 drops of nitric acid. Wash the medicine dropper after every usage. This makes enough aqua regia for about a dozen tests. Only very small amounts of aqua regia should be made at a time. It loses its effectiveness while standing.

Touchstone Test

After an article has been found to be solid gold and not plated, the touchstone method can be used to determine its approximate quality or karat. This is a very old test used by the ancients in a simpler form and perfected by the early alchemists. A jeweler's touchstone is a small stone about two inches square and one half inch thick, cut from black slate or basalt. The surface has been ground to a smoothness that will cause a soft metal, such as gold, to leave a mark or streak when rubbed over it. The test requires a set of standard gold testing needles of various karats from 10 to 24 karats. A set of gold testing needles is usually eight small pieces of brass 1/16 of an inch thick, about two inches long and ¼ of an inch wide at the widest end. The wider end is pierced and strung on a brass ring. These needles are pointed at one end and a tip of standard karat gold has been soldered onto the pointed end. Each needle is marked with the karat of the gold tip.

To make the touchstone test, make a mark on the touchstone with the gold article of unknown karat. On both sides of this mark make similar marks with two of the standard needles. If the unknown is thought to be about 14 karat, rub streaks from the 14 and the 16 karat needles beside the streak of the unknown. For accurate comparison, all marks or streaks on the touchstone should be about the same amount of pressure. A lightly applied mark will dissolve before a heavy one of the same karat. Apply a drop of nitric acid with the tip of the glass stopper across all the streaks. If the unknown changes color and almost disappears, it is 10 karat or less. If there is little change noted from the effects of the nitric acid, the unknown is of higher quality than 10 karat. Wipe the nitric acid off the stone. Next apply a drop or two of aqua regia with the glass stopper and watch the action carefully. If the action of the acid on the unknown does not match one of the first two streaks, try two other needles. When the action of the aqua regia is observed, the color and appearance of the acid treated streaks will correspond when the unknown is of similar gold content to one of the needles. As with any visual comparison test, the accuracy depends upon the judgement and experience of the person making the test. The results shold not be considered as exact as an assay or an accurate chemical determination. With experience, though, a fairly accurate judgement of the karat values can be made. The standard yellow gold test needle set should not be used when attempting to identify the karat of gold of other colors. There are special test sets available for both white and green gold.

The Content of Fine Gold in Karat Alloys

Karat	Percentage of Pure Gold or Parts Per 100	Karat	Percentage of Pure Gold or Parts Per 100
24	100%	16	66.67%
23½	97.92%	15½	64.58%
23	95.83%	15	62.50%
22½	93.75%	14½	60.42%
22	91.67%	14	58.33%
21½	89.58%	13½	56.25%
21	87.50%	13	54.17%
20½	85.42%	12½	52.08%
20	83.33%	12	50.00%
19½	81.25%	11½	47.92%
19	79.17%	11	45.83%
18½	77.08%	10½	43.75%
18	75.00%	10	41.67%
17½	72.92%	9½	39.58%
17	70.83%	9	37.50%
16½	68.75%		

This table is calculated to show the percentage of gold in an article marked with a particular karat mark. The half karat figures are included because many articles of jewelry will actually be one half karat less than the fineness stamped. This is the tollerance the law allows. Most refiners supply goldsmiths and manufacturers with gold sheet and wire of an alloy one half karat less than they intend to stamp the finished article because it is cheaper and it is allowed by law. Also if solder is used in constructing the article of jewelry, it is legal for the article to be one full karat or 4.17% less gold content than the article is stamped. This should be given consideration when buying and selling articles of gold. If an article is not stamped by the maker with a karat value the law or regulations do not apply. Many artisans and craftspeople do not stamp their gold creations.

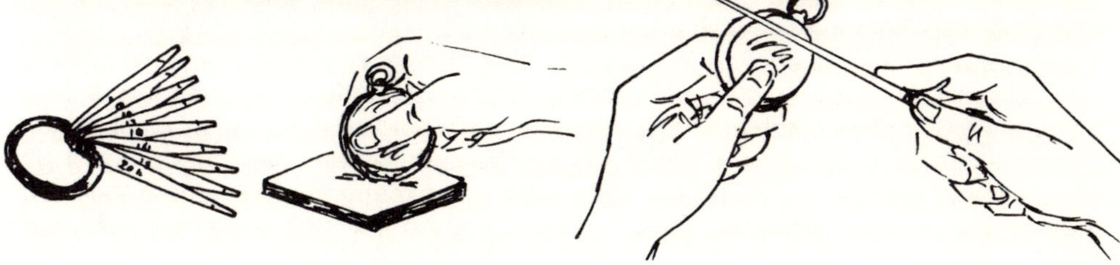

Nitric Acid Test On Gold

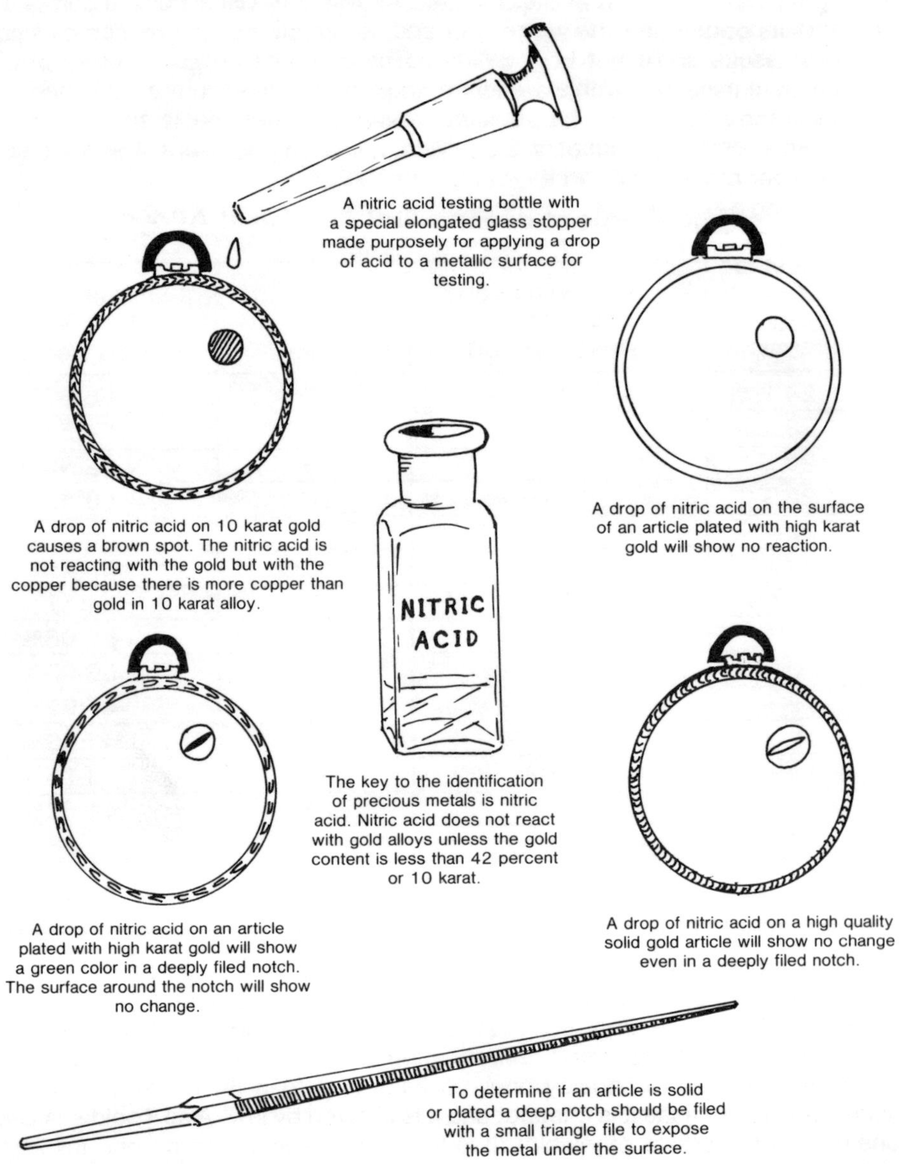

A nitric acid testing bottle with a special elongated glass stopper made purposely for applying a drop of acid to a metallic surface for testing.

A drop of nitric acid on 10 karat gold causes a brown spot. The nitric acid is not reacting with the gold but with the copper because there is more copper than gold in 10 karat alloy.

A drop of nitric acid on the surface of an article plated with high karat gold will show no reaction.

The key to the identification of precious metals is nitric acid. Nitric acid does not react with gold alloys unless the gold content is less than 42 percent or 10 karat.

A drop of nitric acid on an article plated with high karat gold will show a green color in a deeply filed notch. The surface around the notch will show no change.

A drop of nitric acid on a high quality solid gold article will show no change even in a deeply filed notch.

To determine if an article is solid or plated a deep notch should be filed with a small triangle file to expose the metal under the surface.

Gold plated, gold filled or rolled gold articles have a surface finish of gold of high quality, *(but only on the surface)*. To determine if an article is solid or plated, file a deep notch in the article with a small triangle file. The notch must be deep enough to expose the base metal which is usually about the same color as the gold coating. A drop of nitric acid is applied in the notch with the glass tip of the stopper of the nitric acid bottle. If the article is not plated, and is solid gold all the way through no change will be noted. If the article is plated or has a gold coating, the acid will attack the base metal and turn it a green color. The area around the notch, which is covered with gold, will show no change. There are exceptions however: If the core is silver alloy instead of brass or base metal, the notch will show a dark gray color. If the metal exposed in the notch is white, and does not react to the acid, it could be iron or steel. This would rarely occur with an article of jewelry, more probably with such things as gold plated fountain pen points, watch bands, tie slides, etc. A small magnet will quickly identify iron or steel plated articles but some stainless steel is not magnetic. If the article is to be sold as scrap to be melted, a deep file mark is not important. If the article to be tested is to be kept or to be resold, it should not be scarred with a notch. It is sometimes possible, for testing, to file a flat spot along the edge or corner of an article, to expose a small bit of base metal under the plating, instead of filing a deep notch. It is possible to determine if an article is solid gold or gold all the way through, and the approximate karat, by using the specific gravity test. This test requires no marring or filing of the surface, and can be used on any object that can be suspended in water. The specific gravity test is explained in another section in this book.

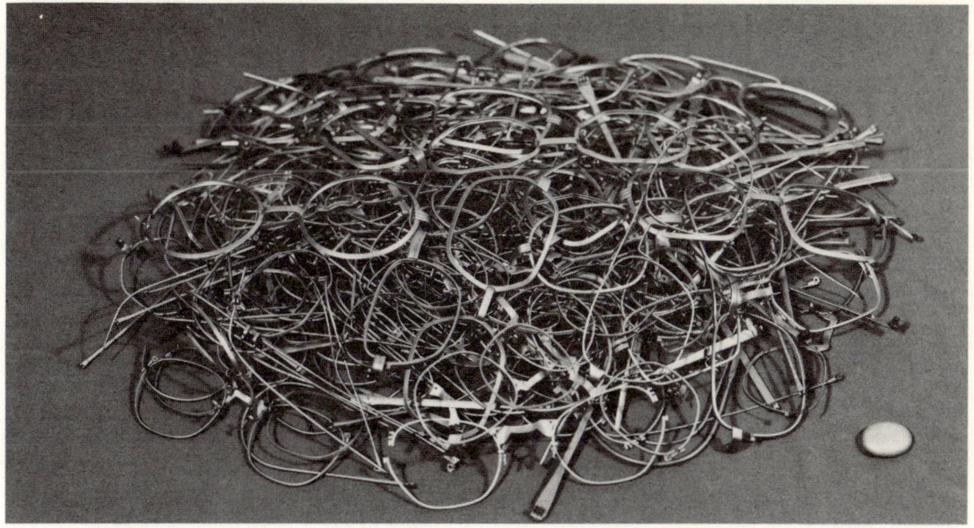

1000 grams (2.2 pounds adv.) of gold plated spectacle frames.

A 25 gram gold button refined from the same weight of spectacle frames.

The pile of gold plated spectacle frames in the photograph weighs 1000 grams or one kilo (2.2 pounds adv.). The 25 gram almost pure gold button in front of the pile was refined from an almost identical pile of frames. About 80% of the 1000 grams of gold plated frames were marked "1/10-10 Karat" or 1/10-12 Karat". Twenty percent were not marked but had a thin gold plate. All showed evidence of wear.

THE AMOUNT OF GOLD ON GOLD PLATED ARTICLES.

Usually the better quality gold plated articles are marked. The markings usually state the quality by weight in proportion to the weight of the article, and the quality or karat of the plate. If an article is marked "1/10 - 12 K. G. P." this means the article is covered or plated with 12 karat gold which makes up one tenth of the weight of the article. There are many different markings in use such as "1/10-10 Karat G.F.", 1/20-12 Karat R.G.P., each discribing the karat and the amount of the gold on the article. To calculate the amount of gold on an article marked "1/10 of 12 karat," first weigh the article, if it weighs one ounce, it would have 1/10 of an ounce of 12 karat gold (when new) covering it. Because 12 karat gold is 50% pure gold the article would have 1/20 of an ounce or one pennyweight of pure gold on it.

Only the thinner and lighter weight items such as spectacle frames and small watch cases will have this amount of gold on them. Many gold plated articles will not be marked with the percentage or the quality of the gold plate, only with the karat, for example "10 Karat Gold Plate" or "12 Karat Gold Plate" or even just "Gold Plate". Frequently articles are marked "Gold Filled" or "Rolled Gold Plate". These articles are made from a special metal plate by soldering one or two layers of gold on to either side or both sides of a base metal plate. This bi-metal or tri-metal plate is rolled thin to form a sheet of metal from which the article is stamped or formed. Almost all other gold plated articles are electroplated after the article is formed.

Large old gold plated pocket watch cases rarely have the weight or the karat values stamped on them because the base metal core is too heavy to be economically covered by one tenth or even one twentieth of its weight in gold. However, these large watch cases usually have a heavy gold plating especially if they are marked with the probable number of years of wear expected or guaranteed. For example there are 10 year, 20 year, and 25 year cases. Precious metal buyers can only give a few dollars an ounce for plated articles. It takes a large quantity to yield enough gold to pay for the refining.

Sometimes an article has gold plate over a sterling silver base or core. A large amount of gold plated jewelry was made during World War II which used a silver base because brass and copper were being used in the production of war materials. The nitric acid test when tried on a piece of gold plated silver-core jewelry can at first be confusing because the silver which is exposed in a filed notch will turn gray instead of green when the acid is applied. Also the article will be a little heavier for it's size than a piece of base metal core jewelry. It should be marked in some way to signify gold plate over sterling.

When calculating the value of gold in an article, and for accurate weight, all the parts that are not gold should be removed, such as springs, screws, cameos and stones. An easy way to remove stones, even diamonds, from a piece of jewelry, without damaging the stones, is by sawing the prongs or bezel almost off with a fine blade in a jewelers saw. Then carefully bend the metal back, thus releasing the stone unharmed. If it is objectionable to remove the stone, the article can be weighed and the weight of the stone estimated and subtracted from the weight of the article.

White Gold

White gold alloys contain the same percentage of gold per karat as the yellow gold alloys, but they also contain enough of some white metal to destroy the yellow color. Usually the white metals used are nickel and palladium. Articles of jewelry that are made of white gold are almost always stamped with the karat value. Testing the quality of a white gold alloy used in an article of jewelry is done the same way as testing yellow gold, using a touch stone and a set of gold testing needles. However a special set of white gold testing needles should be used for accurate results. White gold dental alloys are not marked with a karat value and should be tested to determine their composition and quality. There are a large number of white gold dental alloys, many of them contain gold or platinum and some contain no precious metals at all. The capabilities and equipment necessary for determining the identity of these dental alloys can only be found in large dental laboratories or large precious metal refineries.

Dental Gold

There are probably tons of dental gold tucked away in secret hide-a-way places in millions of American homes. It was put away after being brought from the dentist's office and almost forgotten. The high price of gold should cause these precious mementos to be rescued from jewelry boxes, old coin purses, medicine cabinets or bottom-of-the-drawer clutter and sold to the jeweler or precious metal buyer. Yellow dental golds usually have a very beautiful color and are of very high quality. Most of them are between 18 karat and 22 karat or 75 percent to 90 percent pure gold. This is the visible or exposed yellow gold which is not a hard metal and can be deformed somewhat by the pressures exerted by chewing. Because of this malleability and in order to hold it securely in place, the gold must be soldered to a small piece of another metal or alloy which is harder and not so malleable and that is, in turn, anchored to a tooth or other teeth. Therefore almost all dental gold contains plates and pins. These plates and pins are made up of a variety of metals designed for great strength and hardness and sometimes contain very little, if any, gold. The amount of the backing and pins in a piece of dental gold is usually very small and reduces its overall quality by only a few karats. Most precious metal buyers class yellow dental gold with no visible backing or pins as 18 karat. Others class all dental gold as 16 karat and usually offer to buy it on this basis.

Decisions!

Decisions!

The "white" or silver colored dental alloys are a real problem to identify. Some of these white metals are white gold or platinum-palladium alloys. Usually they are quite heavy and also will react with aqua regia, therefore they can be identified by the touchstone method.

Other white dental alloys may look like platinum but are much too light in weight, they are not magnetic and resist nitric acid just as well as pure gold. Some of these are stainless steel alloys and are not even soluble in aqua regia. The specific gravity test should identify them because they are usually much lighter in weight than the precious metal alloys.

Most precious metal buyers will not buy white colored dental alloys unless they can prove the metal contains gold or platinum.

It will be necessary to clean all the particles of teeth and cement from a piece of dental gold before weighing it. This may be accomplished with a few sharp blows with a hammer while the gold is held on an anvil.

THE APPROXIMATE VALUE IN U.S. DOLLARS OF SOME

Country	Face Value	Date	% of Gold Content	Pure Gold Content Troy Oz.	THE		
					$100	$150	$200
Austria	100 Crown	1915	.900	.9802	98	147	196
Austria	20 Crown	1915	.900	.196	19	29	39
Austria	4 Ducat	1915	.986	.4438	44	66	88
Austria	1 Ducat	1915	.986	.1109	11	16	22
France	100 Francs	1935-36	.900	.193	19	28	38
France	20 Francs	1899-1914	.900	.186	18	27	36
France	100 Francs	1855-1913	.900	.933	93	140	186
Great Britain	Sovereign	1817-19—	.9166	.2354	23	35	47
Hungary	100 Crown	1915	.900	.9802	98	147	196
So. Africa	Krugerand	1967-19—	.9166	1.0	100	150	200
Peru	100 Soles	1950-64	.900	1.354	135	203	270
Peru	50 Soles	1950-64	.900	.677	67	101	135
Peru	20 Soles	1950-64	.900	.256	25	38	51
Mexico	50 Pesos	1947	.900	1.2057	120	180	241
Mexico	20 Pesos	1917-59	.900	.482	48	72	96
Mexico	10 Pesos	1905-59	.900	.241	24	36	48
Mexico	5 Peso	1905-55	.900	.1205	12	18	24
Mexico	2½ Peso	1918-59	.900	.0602	6	9	12
U.S.A.	20 Dollar	1850-1933	.900	.9675	96	145	193
U.S.A.	10 Dollar	1838-1933	.900	.483	48	72	96
U.S.A.	5 Dollar	1837-1929	.900	.241	24	36	48
U.S.A.	2½ Dollar	1837-1929	.900	.120	12	18	24

This chart gives the approximate value in U.S. dollars, of the pure gold content in each of the gold coins listed, depending on the market price of gold. For example, if the market price of gold is $300 an ounce, the gold in the Mexican 50 peso coin will be worth approximately $361.00 U.S.

These values do not reflect any type of premium or commission for buying or selling, only the approximate gold value.

No consideration is shown in the value on the chart for numismatic value for date or rarity. Many of the above listed coins were minted for the express purpose of selling for the bullion value.

Many of the coins listed in this chart are "restrikes" and bear old dates. For example, the Austrian gold restrikes minted after 1915 bear the date 1915. The Mexican 50 peso gold piece was minted from 1921 to 1947 but all 50 peso coins struck in 1947 or after bear the date 1947. The gold coins of the United States all bear the date of the year in which they were struck and there were no restrikes. No U.S. gold coins were minted after 1933. Therefore U.S. gold coins have a genuine historical and numismatic value and usually command, and will always command, a large premium even for the more common varieties. This is in contrast to the Austrian and Mexican restrikes which have only a small premium and could ultimately be worth only their bullion value.

For the thousands of years that mankind has been making and using gold coins, they have been the most positive security anyone could possess. Regardless of what happens to the governments that coin them, gold coins are, and always will be, an international medium of exchange when all else fails.

OF THE MORE COMMON GOLD COINS OF THE WORLD

MARKET PRICE OF GOLD IN DOLLARS PER OUNCE															
$250	$300	$350	$400	$450	$500	$550	$600	$650	$700	$750	$800	$850	$900	$950	$1000
245	294	343	392	441	490	539	588	637	686	735	784	833	882	931	980
49	59	69	78	88	98	108	117	127	137	147	157	166	176	186	196
110	133	155	177	199	221	244	266	288	310	332	355	377	399	421	444
27	33	39	44	50	55	61	66	72	77	83	89	94	100	105	111
48	58	67	77	87	96	106	115	125	135	144	154	164	174	183	193
46	56	65	74	84	93	102	112	121	130	139	149	158	167	176	186
233	280	326	373	419	466	513	560	606	653	700	746	793	840	886	933
59	70	82	94	105	117	129	141	153	164	176	188	200	211	224	235
245	294	343	392	441	490	539	588	637	686	735	784	833	882	931	980
250	300	350	400	450	500	550	600	650	700	750	800	850	900	950	1000
338	406	474	541	609	677	745	812	880	948	1015	1083	1151	1218	1286	1354
169	203	237	270	304	338	372	406	440	473	507	541	575	609	643	677
64	77	89	102	115	128	140	153	166	179	192	204	217	230	243	256
301	361	421	482	542	602	663	723	783	844	904	964	1024	1085	1145	1205
120	144	168	192	216	240	264	288	312	336	360	384	408	433	456	481
60	72	84	96	108	120	132	144	156	168	180	192	205	217	229	241
30	36	42	48	54	60	66	72	78	84	90	96	102	108	114	120
15	18	21	24	27	30	33	36	39	42	45	48	51	54	57	60
241	290	338	387	435	484	532	580	629	677	725	774	822	870	919	967
120	145	169	193	217	241	265	289	314	338	362	386	410	434	458	483
60	72	84	96	108	120	132	144	156	168	180	192	204	216	228	241
30	36	42	48	54	60	66	72	78	84	90	96	102	108	114	120

Since 1974, anyone in the United States can possess as much gold and as many gold coins as he can afford. Gold coins can be brought from abroad, into the United States, and no duties are required on legal tender gold coins.

These coins are seldom sold for the price of their actual gold content, except possibly at banks within the country where they are minted. You can expect to pay a retail premium of 5 to 8 percent over the gold value for the larger than 1 oz. size coins and an escalating premium for the smaller ones, sometimes as much as 50%.

Some countries, such as Canada and South Africa, are minting gold coins that are not stamped with a face value but contain exactly one troy ounce of pure gold and are made expressly for investment. These one ounce coins are an ideal medium for this purpose because their value is published daily as the market price of gold per ounce. Other countries are minting coins with the old face value but they are sold for the market price of the gold content (Mexico, Great Britain, Austria).

The standard fineness of U.S. gold coins, and those of most other countries, is .900 or 21.6 karats. The gold coins minted by Great Britain and the Krugerand of South Africa are .9166 fine or 22 karats. Other countries that use this higher standard are Brazil, Chile, India, Israel, Nepal, Newfoundland, Peru, Portugal and Turkey. Egypt uses a standard of .875 or about 21 karats.

There are probably millions of fake gold coins in existence at the present time. Most of them are made in Beirut or Italy. They are almost perfect copies of rare coins and are sold to tourists or uninformed collectors at a price less than the numismatic value but well above the value of the gold content. A knowledgeable coin dealer should be able to identify these fakes. Fortunately they usually contain as much fine gold as the coins they duplicate and upon identification, should be melted and sold as bullion.

FLUCTUATIONS IN THE PRICE OF GOLD
1970 - 1980
(In Dollars Per Ounce)

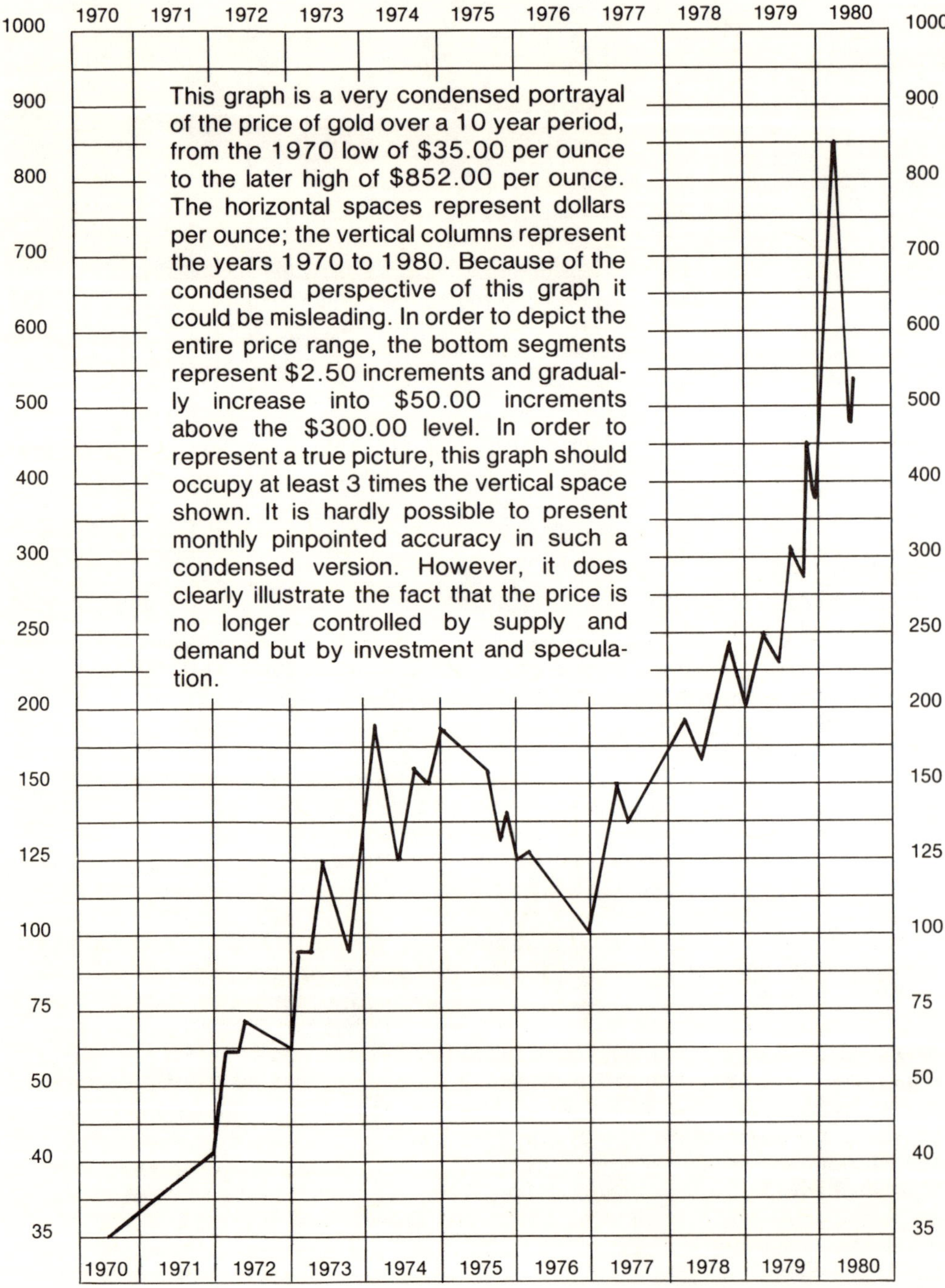

FLUCTUATIONS IN THE PRICE OF SILVER 1970 - 1980

(In Dollars Per Ounce)

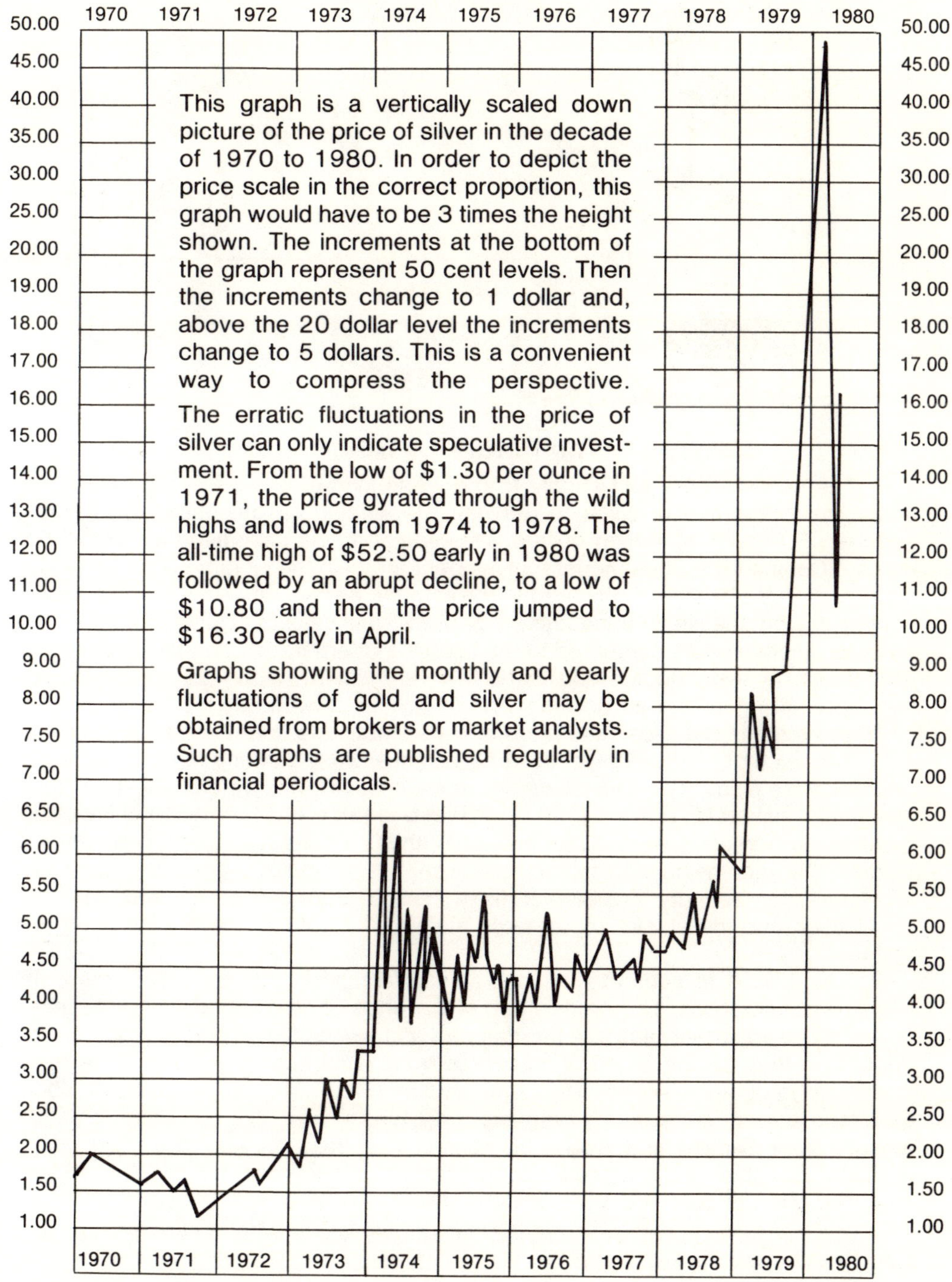

Anything goes — nothing is sacred.

The Future Of Gold

The newly mined gold supply to the world market is quite small, usually not much more than five or six tons per day. Depending on the price, the value is between 50 and 100 million dollars. This seems to be a tremendous amount of money. Compared to the amount that is spent for oil in the oil rich countries, this is just a drop in the bucket. Compared to the amount the federal government is spending per day on seemingly unnecessary projects, this amount is just "peanuts". It is an insignificant amount compared to the billions that are traded on Wall Street every day. It can be easily understood that a few very wealthy individuals or companies buying or selling millions of dollars worth of gold or silver a day for the purpose of profit investment can cause the price of these metals to fluctuate in their favor.

The price of gold and silver has been gradually rising during the last decade to unpredictable highs. The daily or weekly ups and downs clearly indicate speculative investment for profit. But the gradual increase in price points to several important factors: the increasing demand for the use of these metals in industry, dentistry and jewelry; the increasing demand for these metals, especially gold, by millions of individuals for security investment as a hedge against inflation and currency devaluation and the stockpiling of large amounts of gold by money interests and nations.

It is difficult to separate the real or natural market demands from the profit investment influences. The gradual rise of the price of gold and silver is influenced by the more important fact that the demand and use of these metals is increasing faster than they can be mined or recovered from secondary sources such as scrap metal or photographic waste. The consumption in the United States alone is about 200 tons a year. We produce only about 40 tons a year. We buy about an equal amount from Canada and over 100 tons a year from foreign countries. If the demand continues and the production decreases, which it is doing at the present time, the price of gold and silver will gradually continue to rise.

It is difficult in uncertain economic times to decide whether to keep or sell your gold or silver. As the price of these metals increases it is tempting to "cash in" and receive the large amount of "paper" money which is losing it's value as fast as the gold or silver is increasing in value.

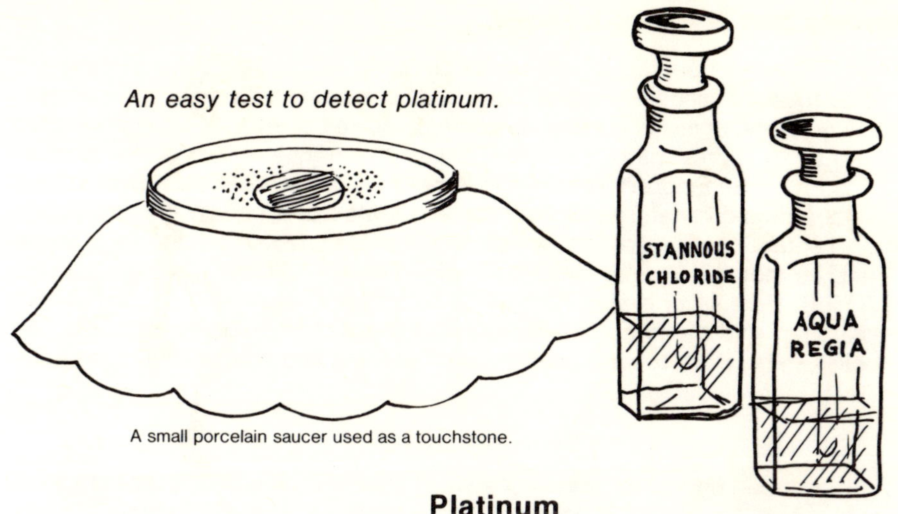

An easy test to detect platinum.

A small porcelain saucer used as a touchstone.

Platinum

There are several white metals used in jewelry that could be mistaken for platinum and like platinum, they are not affected by nitric acid. The more common are stainless steel and white gold. Stainless steel can be identified by its light weight (specific gravity test) and also most stainless steel alloys will react to a drop of hydrochloric acid which slowly forms tiny bubbles on the surface of the metal.

White gold and platinum alloys can be confused when being tested on the touchstone, both are dissolved by aqua regia, white gold dissolves rapidly and platinum slowly. This should not be considered a conclusive test. Two samples of white gold might dissolve differently. One might dissolve more rapidly than the other because of less gold content.

Articles of platinum jewelry are usually marked "Platinum" or "Pt.". The standard test for platinum or platinum-palladium alloys, is the stannous chloride test. Stannous chloride test solution can be made by dissolving about a penny-weight or less of stannous chloride crystals or powder in a one-half ounce acid testing bottle three-fourths full of water, then adding 25 drops of hydrochloric acid. This makes a milky solution which will lose its strength in a few days. A small piece of pure tin the size of a match head can be added to lengthen the shelf life of this solution.

Instead of the black touch plate, a white one should be used so the results, which show colors, can be seen more easily. The bottom of a small white porcelain saucer or a broken piece of white porcelain can be used. A dull spot the size of a fifty cent piece should be rubbed on the glaze using a piece of emery cloth. The spot should be large enough on which to make streaks or marks with the metals being tested. Rub the metal on the roughened spot, making a heavy streak, and dissolve the mark with aqua regia. When cold, platinum dissolves slowly in aqua regia, more rapidly when heated. This can be done by warming the porcelain saucer over an electric bulb for a few minutes. If the acid evaporates, add another drop or two of aqua regia. After the metallic streak is dissolved add a drop or two of stannous chloride solution. If the metal in the streak was platinum, at first a deep yellow or brown color will form. If the streak was very rich in platinum, the color will turn almost black. If palladium is present with the platinum the color will turn blue-green.

If the streak was gold and stannous chloride solution is added, the first intense dark color will be purple and then black. A tiny amount of gold will turn the solution purple, a little more will turn the solution black. The solution can be diluted with several drops of water to bring the color of the solution back to purple. This is a very sensitive test for detecting gold.

Silver

Silver, the most beautiful of the white precious metals has been sought and treasured by mankind since very early times. It is considered the feminine companion metal to gold, the masculine metal. In some primitive societies, gold represented the sun and silver represented the moon. Silver is non-magnetic and, like gold, is one of more than one hundred elements that make up the planet Earth. Unlike gold, silver is seldom found in nature in a pure metallic state but is usually in combination with other elements as an ore which has to be processed or refined in order to obtain the pure metal. It is only about one half as heavy as gold, having a specific gravity of 10.5. Pure silver is about the same hardness as pure gold (2½-3). It cannot be scratched with the fingernail (as can pure lead) but can be easily scratched with a pocket knife. Pure silver is a little less malleable and ductile than pure gold.

Sterling Silver

When stamped on a silver article, the word "Sterling" is one of the best known and most respected markings in use on a precious metal. The word came into prominence in the 12th century when five free cities in eastern Germany joined together to form a trade league and issued their own coins. These reliable, high quality coins were much in demand as a medium of exchange even outside the league and became known as the coins of the "Easterlings". The name was shortened to "Sterling" and now is used on silver as a stamp of high quality.

This mark designates that the article on which it is stamped is made of 92.5 percent pure silver and 7.5 percent copper. The copper is added to give strength and better wearing quality because pure silver is too soft for ordinary use.

The sterling mark or stamp is the standard marking used in the United States. Very rarely will the mark .925 be used. Seldom will the sterling mark be found on articles made in Great Britain, unless they are made especially for export. Sometimes the quality marks and percentage of fineness marks are required (.925 or .9584) on articles imported into Great Britain.

Coin Silver

Coin silver is 90% silver and 10% copper. This alloy is harder and tougher, consequently more suitable than sterling for use in coins. This small percentage of difference, 2½% more copper, improves the wearing qualities needed in coins but does not change the appearance or color. It is very difficult to distinguish sterling silver from coin silver, even for an experienced person. The amount of pure silver in sterling and coin differs by only 25 parts per 1000. This small variation in the amount of silver makes distinguishing between the two very difficult using simple tests. Some expert silver buyers can tell the difference by using known pieces of coin and sterling and then comparing the slight color changes with the color change of an unknown example. When drops of nitric acid are applied to all three at the same time and all three are exposed to direct sunlight, the sunlight causes the silver nitrate made by the acid to darken slightly faster and darker on the sterling than on the coin silver. If a small amount of another metal, such as lead or zinc, is in the copper alloy making up the unknown, it will make this test inconclusive. When this test is tried on silver alloys of lesser quality, the lower the quantity of silver in the alloy the greener will be the color produced by the nitric acid.

Because of the high price of silver, there are many fine old pieces of unmarked coin silver tableware being offered for sale today. These "Early American" or "Colonial" handmade spoons and forks were made from melted British or Mexican coins and from any other silver that was available. Usually they were unmarked but a few bore the maker's name and sometimes the word "Coin", "Dollar" or "D". The only way to determine the quality is by testing with acid or dichromate solution.

One of the interesting and, at times, annoying characteristics of silver is it's ability to tarnish and form a patina on the surface. This is a dark or black coating caused by the sulfur gases in the air forming silver sulfide on the surface (not oxidation). This darkening outlines and accentuates any stamping or incised portion of a design into the surface of the metal and causes the higher parts of the surface to be highlighted when rubbed or polished. This creates a most pleasing effect on coins, medals, jewelry or sculpture.

It is also this ability of silver compounds to darken on exposure to light that makes its use in photography invaluable. Tremendous quantities of silver are used in the production of photographic films and prints. Many years of research have been spent with little success, trying to find a substitute for silver in this industry.

The Composition Of Silver Alloys

How silver is usually marked and what these marks mean.

These are the common alloys of silver, the way they are marked and the composition or percentage of fine silver and base metal of which they are made.

The Alloy	Mark	Fine Silver	Alloy
Fine or pure silver, usually bars or ingots	.999+	99.9+%	trace
Britannia silver, highest grade used in Great Britain — very soft	Hallmark figure of Lady Britannia	95.84%	4.16%
Sterling silver	Sterling	92.5%	7.5%
Coin	Coin or Coin silver	90%	10%
Low quality European silver	.800	80%	20%
Low quality silver alloy, usually novelty items made for the tourist trade in Europe, India, Mexico and the Orient.	"Silver" or no mark	.720% to .500%	28% to 50%

Articles stamped "Sterling Silver" should assay .925 fine. However, the regulations require them to be only .921. Fine sterling tableware is usually within this tolerance if no solder is used. Sterling silver articles assembled with solder are required to assay only .915 pure silver. Silver articles stamped "Coin Silver" should assay .900 fine. They are only required to be .896 fine or, if solder is used, only .890.

The quality or percentage of silver in coins is listed in the coin charts elsewhere in this book.

When the term "solid silver" is used in this book, it refers to an article that is silver all the way through, regardless of composition or quality. It can be either pure, sterling, 80% or 50% silver or an alloy that is mostly silver. This is intended to distinguish the solid articles from the coated or plated articles. The alloys of very low silver content are sometimes referred to as alloy silver.

How To Determine If An Article Is Solid Silver Or Silver Plate

It is reasonable to assume that articles marked "Sterling" are solid silver and are not plated. Most precious metal buyers accept this mark without testing unless the look or the feel of an article is questionable. Most sterling silver articles, especially spoons and forks have a certain feel and springiness that plated ware does not have, regardless of the weight or thickness.

A drop of nitric acid on the surface of a silver colored article will quickly determine if the metal is silver. Nitric acid reacts very quickly with silver. At first, it causes a dark spot, then it leaves an etched gray area after the acid is wiped off. This test will determine that the surface of the article is silver. To find out if the article is solid silver a notch should be filed in the surface, deep enough to penetrate the silver coating and expose the base metal core. A drop of nitric acid is applied to the notch filed in the surface. If the article is solid silver, the metal in the notch and surface around the notch will turn gray from the action of the acid. If the article is plated, and the notch has been filed deep enough, the area around the notch will turn gray but the exposed base metal in the notch will turn green. This is because the nitric acid turns the copper in the base metal into copper nitrate which is green.

If the article is to be sold for scrap metal, a deep notch filed in the surface is not important. If the article is too valuable to be scarred or filed, the specific gravity test will quickly determine if the article is solid or if it is plated. Unless it is positive the article will be melted as scrap metal, nitric acid should be applied to an article in a spot that will be as inconspicuous as possible. Great care should be used when testing with nitric acid as serious burns can result when the acid comes in contact with the skin or eyes. When applied to a metal article, the acid should be wiped off with a cloth or paper towel immediately after the results are decided. Deep etching to the metal can result and a careless touch of the acid can cause a burn.

Another method for testing silver articles is to use a dichromate solution. This is an easy solution to make and simple to use. Dissolve about 1/8 teaspoon of potassium dichromate in about 1/4 ounce of nitric acid. Simply add the dichromate to the nitric acid in one of the glass "acid testing" bottles until the solution is a deep burgundy color. When a drop of this solution is put on an article of silver, the color changes to a bright red. The nitric acid dissolves some of the silver and it is changed to bright red silver dichromate. If the dichromate solution is applied to a silver plated article with a notch filed through to the base metal, it colors the base metal green, the same as in the regular nitric acid test. The dichromate test is considered to be the most positive test for silver. No other silver colored metal gives this red indication; lead shows a yellow color; pewter and Britannia metal turn dark then almost black; iron bubbles and turns dark; white gold, stainless steel and platinum show no change.

Silver tableware, both solid and plated, made in Great Britain and the European countries, is frequently marked with "Hall Marks", one of which is the quality mark. Seldom is the ware marked with the numerical fineness of .925, the word "Sterling" or the word "plated". Because of the many different kinds, the quality marks are very difficult to interpret except by an expert.

Hallmarks

Quality Marks

Date Letters

Nitric Acid On Sterling Or Solid Silver

A small drop of nitric acid on the surface of a solid silver article will cause a gray spot etched into the metal. The same color will be observed in the file mark. No other silver white metal will give this reaction. "Nickel-silver" or "German-silver" will turn green. Other white metals that resemble silver such as high quality white gold, stainless steel or platinum will show no reaction.

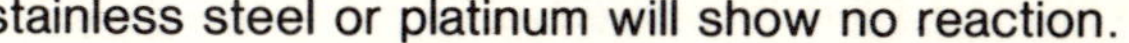

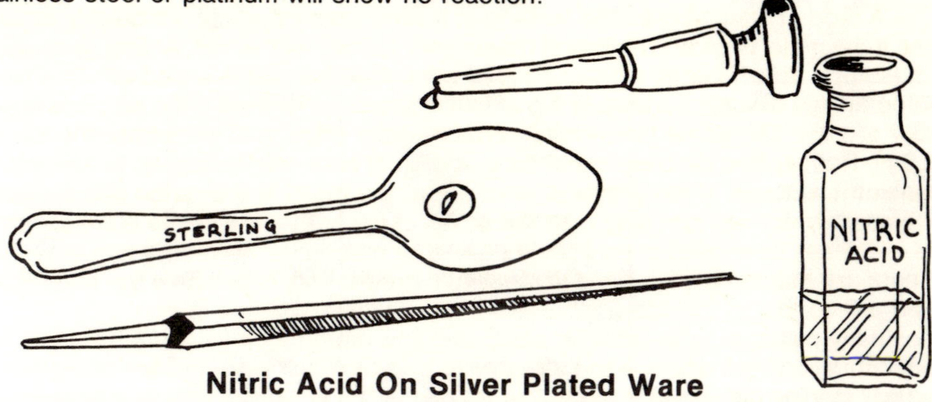

Nitric Acid On Silver Plated Ware

A drop of nitric acid in a deep file mark on a piece of silver plated ware will show a green color in the file mark if the base metal core is exposed. The silver coating around the file mark will be gray.

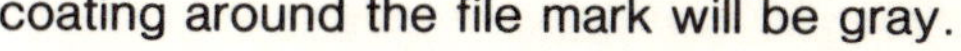

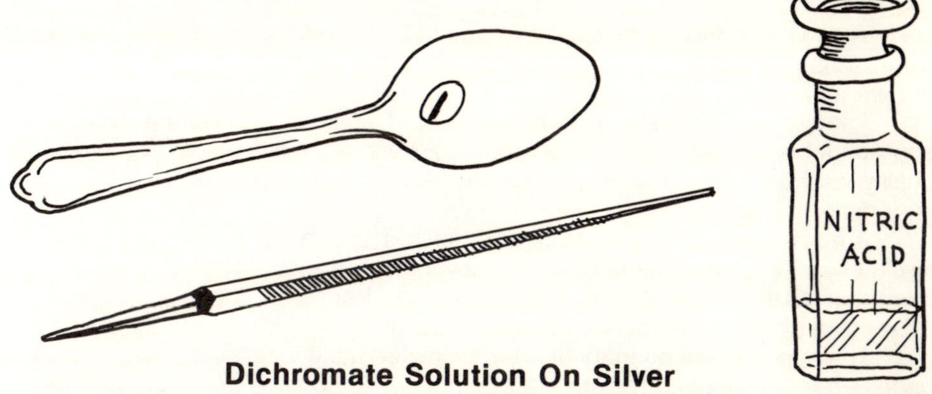

Dichromate Solution On Silver

A drop of dichromate-nitric acid test solution on silver will make a bright red spot which can be wiped off to leave an etched gray surface. No other white metal but silver will show this bright red color. A file mark on plated ware will turn green the same as with the nitric acid test.

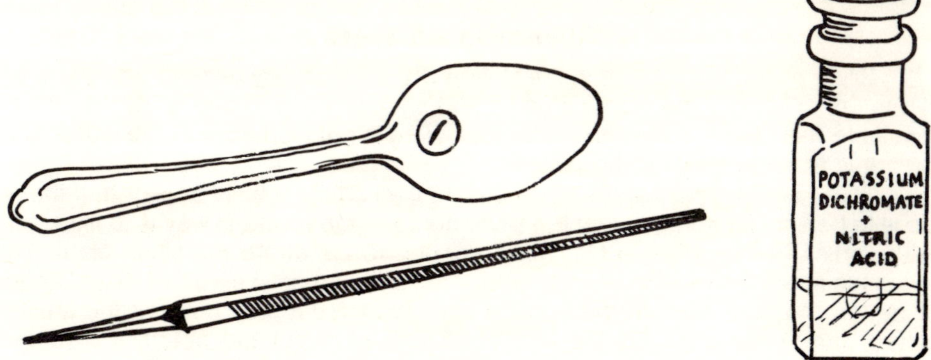

The Amount of Silver on Silver Plated Table Ware

The silver plated ware found today will not be marked "Sterling" but will have one or more of these marks: the name of the company, a mark designating the thickness of the coating and usually the word "Plate" or "Plated".

Company Name	Quality Marks	Thickness Of Plate	Troy Ounce per square foot of surface
International Silver	A1 or standard	.00028 inch	.222
IS	A1 or A1X or		to
1847 Rogers Bros.	extra		.278
Oneida-Tudor	AA		
Community	Double or XX	from .00042	.333
	Triple or XXX	to	to
There are many others	Quadruple or XXXX	.00111+ inch	.888
and unless the			
company name is			
accompanied by the	Extra heavy	from .00125	1.00
word "sterling" it		to	
is probably plated ware.		.0018 inch	

The manufacturers claim the average thickness of the silver plate on tableware varies from .00028 to .0018 of an inch thick, depending on the type and quality of the ware.

Another way of designating the amount of silver on tableware is in ounces per square foot of surface. This varies from .222 ounces per square foot for the lower quality ware to over 1 ounce per square foot for the heavier plate. This would be very difficult to calculate.

A more interesting way to classify plated spoons and forks (and one which could be more easily checked) is the designation of the amount of silver per gross pieces or 144 pieces. This is estimated to be from 2 to 9 ounces of pure silver depending on the manufacturer and the quality.

The approximate quantity of silver on plated tableware varies from about 2 to 4 ounces per 144 pieces for the standard plate, to about 8 or 10 ounces per 144 pieces for the thickest plate. This is an average for slightly worn spoons and forks of varying sizes.

Silver Plated Ware

Silver plated ware made in the United States is usually marked with the words "Plate" or "Plated", the maker's name and sometimes with one or two marks that resemble Hall Marks. If modern silver tableware is not marked with the word "Sterling", it is almost certain to be plated.

There are two easy ways to determine if an article is solid or plated. If the article is to be sold for scrap or just the silver content, the quickest way is to file a notch in the article with a triangle file, deep enough to penetrate the silver plate. Then touch the notch with a drop of nitric acid. If the article is solid silver, of .900 fineness or better, the acid will turn the surface and the notch a gray color. If the article is plated, the surface around the notch will turn gray but the notch, if filed deep enough to expose the core, will turn green because the nitric acid turns the copper in the base metal core into copper nitrate which is green.

The second method to use is the specific gravity method, especially if the article is too valuable to be filed and small enough to be weighed in a receptacle of water and then in air. The weight of the object divided by the loss of weight when in water is its specific gravity, the resulting number will quickly determine if the article is solid silver or plated.

144 Silver Plated Spoons and Forks

144 Spoons and Forks with the plating removed and the two ounce button of silver from these articles.

Two almost identical piles of silver spoons and forks are shown in the above photograph. The articles in the pile on the left still have the silver plating. The silver has been removed from the articles in the pile on the right and the silver from the articles is shown in the button in front of the pile. The silver was removed by using a special combination of acids, and then refined and melted into a lump of pure silver. It was slightly less than two ounces in weight. This process of removing silver from plated ware is expensive and time consuming. It would be profitable only when the price of silver is above 30 dollars per ounce and then only if hundreds of pounds of silver plate was processed at one time.

Usually worn and damaged silver plate is sold as junk metal to be melted, with other base metals, into large slabs for refining electrolytically and the silver is then recovered as a by-product.

Archimedes discovering the principles of specific gravity.

Specific Gravity

"Some things are heavier than others" is a very simple explanation of specific gravity. A gold ring is almost six times as heavy as an aluminum ring of the same size and dimensions. Sometimes this is referred to as "heft". This is due to the difference in the density of the two metals. A teaspoon full of pure gold is over 19 times as heavy as a teaspoon full of water. Water with the density or specific gravity of 1.0 is used as the basis of comparison. The specific gravity of a substance is the number of times that substance is as heavy as water. Pure gold is considered to have the specific gravity of 19.3, or gold is 19.3 times heavier than water, volume for volume.

Archimedes (250 B.C.) was probably the first to calculate the difference in density of one substance compared to another. The emperor wanted to know if his crown, which was supposed to be pure gold, did not in fact contain some silver. One day when Archimedes stepped into his bath and caused the water to overflow, he perceived that, by using this same method, the excess of bulk caused by the introduction of a lighter alloying metal could be measured by putting the crown and equal weights of gold and silver into separate bowls of water and measuring the overflow from each bowl. What Archimedes discovered was a way to calculate the ratio between the weight and the bulk of an article. This ratio is called the density or specific gravity. "Specific" because each pure elemental material has its own unique and specific ratio and "gravity", of course, refers to weight.

The identity of most precious metals can be determined with the touch of a drop of acid. Because there is the possibility of a precious metal coating covering a base metal core, the composition of an article may not be what it appears on the surface. In acid testing, the only way to determine what is under the surface is to file a deep notch and apply acid. This is of no importance if the article is to be sold for scrap metal. If the article has artistic or antique value, a notch, or even the etching left by a drop of acid, would probably ruin its value. In most cases the composition of an article can be determined very easily if it is small enough to be submerged in a container of water while it is being weighed. Items of jewelry with stones or parts that are known not to be precious metal, that cannot be easily removed, should not be considered as subjects for a specific gravity test. Only the weight of the metal is necessary. Even small bubbles of air adhering to the article can cause an incorrect result. The smaller the article the less accurate the determination will be. For very

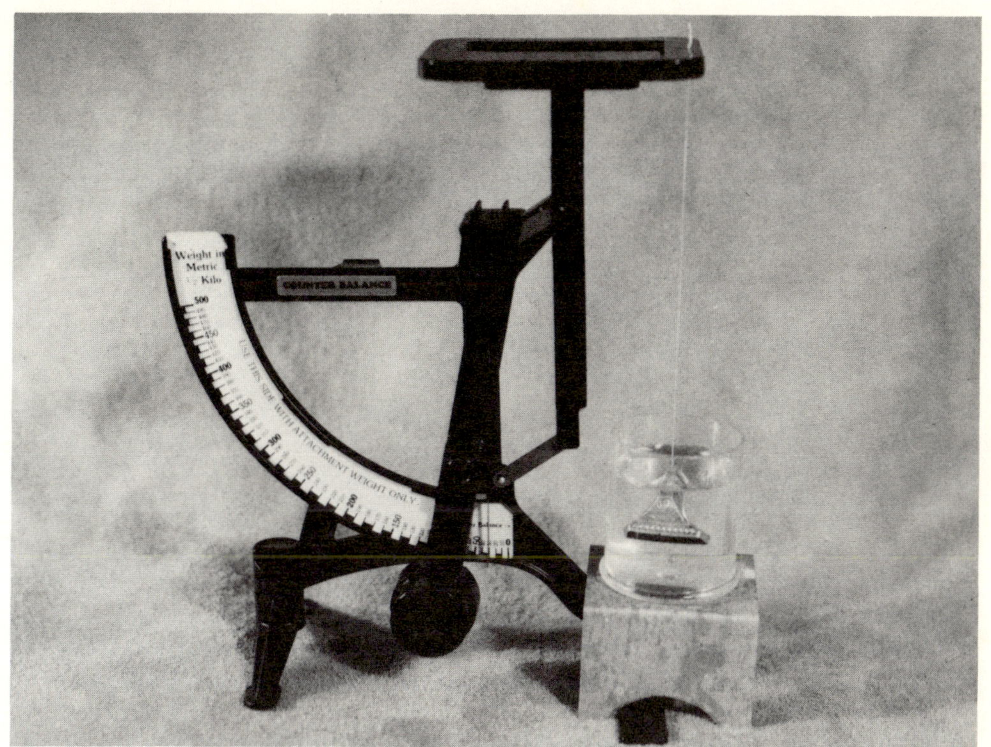

small articles a sensitive laboratory balance should be used to get extreme accuracy. In most cases a small, inexpensive, accurate balance, like the one shown above, is all that is necessary to obtain a fairly close specific gravity determination. The weighing should be done with grams and milligrams because the calculations are much easier when using the metric system. However weighing in ounces can be done just as accurately.

The easiest way to weigh an article, both in air and in water, is to suspend it by a thin thread in a position so it can be weighed in water in a small container after it has been weighed in air. The smaller the thread holding the article the better the accuracy.

A good example is the ring, with the stone removed, in the photograph above. It is suspended by a tiny nylon thread which is fastened through a small hole in the balance pan. It was weighed while suspended in air and found to weigh 33 grams. A small plastic container (pill bottle) was filled with water and carefully placed on a bench made of scrap aluminum sheet. The ring was suspended in a position not touching the sides or bottom of the container and weighed under water. It was found to weigh 30 grams. To find the specific gravity, subtract the weight in water (30 grams) from the weight in air (33 grams) which gives a loss of weight in water of 3 grams. The weight in air (33 grams) divided by the loss of weight in water (3 grams) gives the figure 11.0. This is the specific gravity of the ring. The ring was thought to be gold but did not feel quite heavy enough and it was not stamped with a karat mark. Nitric acid had no effect on the surface. The owner did not want the ring filed. The stone was removed, the ring weighed and the specific gravity found to be 11.0. This figure is slightly higher than sterling (10.4). From this information it was decided that the ring was sterling silver with a heavy gold plating. The gold plating would account for the specific gravity being slightly higher than for sterling silver.

The term "karat" is a very old and convenient way to designate the amount of gold in an alloy or an article. A karat is not a unit of weight such as the gram or the pennyweight but only an expression of the proportion of gold to the amount of alloy. The karat should be considered as a figure of the percentage of purity, 24 karat means pure gold, 12 karat gold means 50% gold and 50% base metal alloy.

An ancient Egyptian gathering seeds from the carob tree to use as weights.

The word "Karat" possibly originated from the use of the seeds of the carob or locust tree. These seeds were used as weights to weigh precious metals in early times. The seeds were quite common, heavy for their size and fairly uniform in size and weight. The fully formed seeds, taken from the center of the bean or seed pod, were accepted as a basis of weight throughout the Mediterranean and Egyptian trading communities. In Roman times the karat was used as a measure of weight and fineness, representing the Roman weight siliqua, or carat, which was 1/24th the weight of the golden coins or Solidus of Constantine. In time it was used to designate the purity or fineness of gold. In the U.S.A. the form "karat" (with a k) is used for the ratio of gold in an alloy. Outside the U.S.A. the form "carat," as well as other spellings, is used for both meanings. The word carat (with a c) is the name for a weight (200 milligrams) used for weighing precious and semiprecious stones.

A Comparison Of A Few Of The Most Useful Weights Used In Weighing Precious Metals

Troy	Metric Equivalent	Avoirdupois
1 grain =	.648 grams	= 1 grain
1 pennyweight = 24 grains = 1.5517 grams =	1.5517 grams	= 24 grains
1 troy ounce = 480 grains = 20 pennyweight =	31.103 grams 28.35 grams	= 1 avoirdupois ounce = 437.5 grains

The Value Of Silver In Coins

The charts on the two following pages show the basic silver values for the more common silver coins of the United States, Canada and Mexico. The values shown are dollar values per troy ounce for the quantity of silver each newly minted coin theoretically contains when the market price of silver is at a certain price level. Circulated and worn coins should be considered of less value, depending on the amount of wear they show. These charts give the values of pure silver in these coins when the market price of silver is between 10 and 50 dollars per ounce.

These charts may also be used to calculate the value of silver in these coins even if the market price of silver is lower than 10 dollars per ounce or over 50 dollars per ounce. If silver is between 5 and 10 dollars an ounce, the market price values between 10 and 20 dollars can be used by taking one half of that value listed. For example, to check the silver value of the U.S. dollar at the market price of 7 dollars an ounce, take one half of the 14 dollar market price of the coin, which is one half of $10.82 or $5.41.

If the market price rises to over 50 dollars per ounce, add to the $50.00 figure the necessary amount from the chart. For example, if silver is 56 dollars per ounce market price and you would like to find the value of pure silver in a U.S. dollar at this price, add the 50 dollar value ($38.67) plus one half of the 12 dollar market price figure which is $9.28 divided by 2 or $4.64. Add this figure (which is the value of the market price figure at 6 dollars an ounce) to the 50 dollar figure of $38.67. Therefore, the value of the pure silver in a U.S. dollar at the 56 dollar market price is $38.67 plus $4.64, which equals $43.31.

If the price goes into the 100 to 500 dollar range, the entire chart can be converted into this price range by moving the decimal point one place to the right. For example, if the market price of silver is $250 per ounce and you would like to find the pure silver value in the U.S. dollar, use the 25 dollar column which shows the value of the U.S. dollar to be $19.33, move the decimal one place to the right in both figures and the value at 250 dollars per ounce will be $193.30.

Silver coins should be checked for their possibly greater numismatic value or rarity before selling them for only the silver metal value.

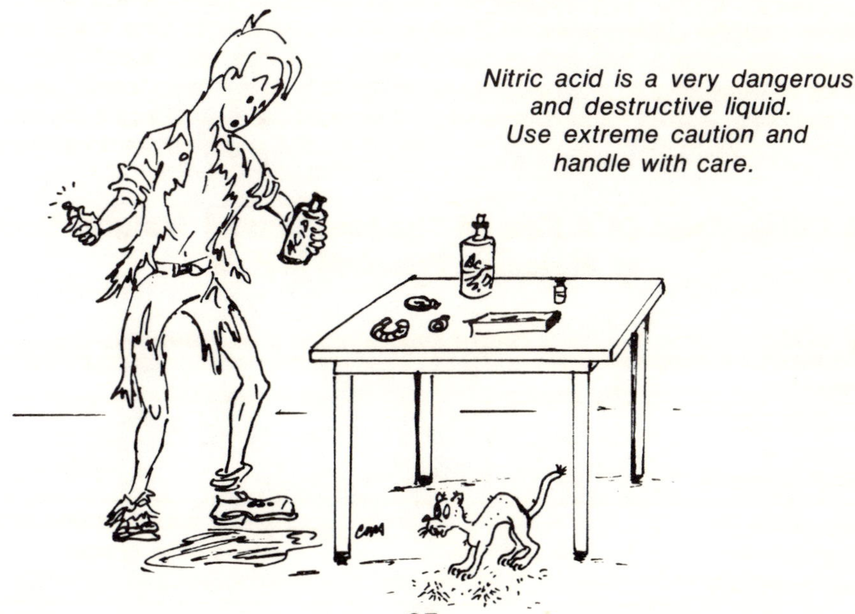

SILVER VALUE CHART FOR THE MORE COMMON SILVER COINS OF THE UNITED STATES AND CANADA

	% Silv.	THE MARKET PRICE OF SILVER IN DOLLARS PER OUNCE																				
		$10	$11	$12	$13	$14	$15	$16	$17	$18	$19	$20	$21	$22	$23	$24	$25	$26	$27	$28	$29	$30
U.S. 5¢ War Nickels	.350	.57	.62	.68	.73	.79	.84	.90	.96	1.01	1.07	1.13	1.18	1.23	1.29	1.35	1.41	1.46	1.52	1.58	1.63	1.69
U.S. 50¢ Clad '65-'69	.400	1.48	1.63	1.77	1.92	2.07	2.22	2.37	2.51	2.66	2.81	2.99	3.11	3.25	3.40	3.55	3.70	3.85	3.99	4.14	4.29	4.44
U.S. 10¢	.900	.73	.81	.88	.95	1.02	1.09	1.16	1.24	1.31	1.39	1.45	1.52	1.59	1.66	1.73	1.80	1.88	1.95	2.02	2.09	2.16
U.S. 25¢	.900	1.80	1.98	2.16	2.34	2.52	2.70	2.88	3.06	3.25	3.43	3.60	3.78	3.96	4.14	4.32	4.50	4.68	4.86	5.04	5.22	5.40
U.S. 50¢	.900	3.62	3.99	4.35	4.71	5.07	5.43	5.79	6.15	6.51	6.87	7.22	7.58	7.94	8.30	8.66	9.02	9.38	9.75	10.10	10.47	10.83
U.S. 1.00	.900	7.73	8.50	9.28	10.05	10.82	11.60	12.37	13.14	13.90	14.70	15.47	16.24	17.01	17.79	18.56	19.33	20.10	20.88	21.65	22.43	23.20
Canadian 10¢	.800	.59	.65	.71	.77	.83	.90	.96	1.02	1.08	1.14	1.19	1.25	1.30	1.38	1.44	1.49	1.56	1.62	1.68	1.74	1.79
Canadian 25¢	.800	1.49	1.64	1.79	1.93	2.08	2.25	2.39	2.55	2.69	2.85	2.99	3.15	3.29	3.45	3.59	3.75	3.89	4.05	4.19	4.35	4.49
Canadian 50¢	.800	2.99	3.28	3.58	3.89	4.19	4.49	4.79	5.09	5.39	5.59	6.00	6.29	6.59	6.90	7.19	7.50	7.79	8.09	8.39	8.69	8.99
Canadian 1.00	.800	6.00	6.60	7.20	7.80	8.40	8.99	9.59	10.20	10.80	11.40	12.00	12.60	13.20	13.80	14.40	15.00	15.60	16.20	16.80	17.40	18.00
Canadian 10¢	.500	.37	.41	.44	.48	.52	.56	.59	.63	.67	.70	.75	.79	.82	.86	.90	.94	.97	1.01	1.05	1.08	1.12
Canadian 25¢	.500	.93	1.02	1.12	1.21	1.31	1.41	1.50	1.59	1.68	1.78	1.87	1.97	2.06	2.15	2.25	2.34	2.43	2.53	2.62	2.72	2.81

	% Silv.	THE MARKET PRICE OF SILVER IN DOLLARS PER OUNCE																			
		$31	$32	$33	$34	$35	$36	$37	$38	$39	$40	$41	$42	$43	$44	$45	$46	$47	$48	$49	$50
U.S. 5¢ War Nickels	.350	1.74	1.80	1.86	1.91	1.97	2.02	2.08	2.13	2.19	2.24	2.30	2.35	2.41	2.47	2.52	2.58	2.64	2.69	2.75	2.80
U.S. 50¢ Clad '65-'69	.400	4.58	4.73	4.88	5.03	5.18	5.33	5.48	5.63	5.78	5.93	6.07	6.22	6.37	6.51	6.66	6.81	6.96	7.11	7.25	7.40
U.S. 10¢	.900	2.24	2.31	2.38	2.45	2.52	2.59	2.66	2.73	2.80	2.89	2.96	3.03	3.10	3.17	3.24	3.32	3.39	3.46	3.53	3.60
U.S. 25¢	.900	5.58	5.76	5.94	6.12	6.30	6.48	6.66	6.84	7.02	7.20	7.38	7.56	7.74	7.92	8.10	8.28	8.46	8.64	8.82	9.00
U.S. 50¢	.900	11.19	11.55	11.91	12.27	12.63	12.99	13.35	13.72	14.08	14.44	14.80	15.16	15.52	15.88	16.24	16.60	16.96	17.33	17.69	18.05
U.S. 1.00	.900	24.00	24.80	25.50	26.20	27.06	27.84	28.61	29.38	30.15	30.93	31.70	32.48	33.25	34.02	34.80	35.57	36.35	37.12	37.89	38.67
Canadian 10¢	.800	1.86	1.92	1.98	2.04	2.09	2.14	2.20	2.26	2.32	2.38	2.44	2.50	2.56	2.62	2.68	2.74	2.79	2.85	2.91	2.97
Canadian 25¢	.800	4.65	4.79	4.95	5.09	5.25	5.36	5.53	5.68	5.81	6.00	6.15	6.30	6.45	6.60	6.75	6.90	7.05	7.20	7.35	7.50
Canadian 50¢	.800	9.29	9.59	9.89	10.20	10.50	10.78	11.06	11.38	11.66	11.98	12.26	12.56	12.85	13.15	13.45	13.75	14.05	14.35	14.65	14.95
Canadian 1.00	.800	18.60	19.20	19.80	20.40	21.10	21.60	22.20	22.80	23.40	24.00	24.60	25.20	25.80	26.40	27.00	27.60	28.20	28.80	29.40	30.00
Canadian 10¢	.500	1.16	1.19	1.24	1.27	1.31	1.33	1.36	1.40	1.44	1.48	1.52	1.56	1.60	1.63	1.67	1.71	1.75	1.78	1.82	1.86
Canadian 25¢	.500	2.91	2.99	3.09	3.18	3.28	3.36	3.46	3.55	3.65	3.74	3.83	3.92	4.02	4.11	4.20	4.30	4.39	4.48	4.58	4.67

The values listed in the charts on these two pages reflect only the approximate value of the silver contained in a standard newly minted coin. No allowance has been made for slight variation in silver content or for normal circulation wear. Most of the coins were minted in such large quantities they will have little or no special numismatic value.

SILVER VALUE CHART FOR THE MORE COMMON SILVER COINS OF MEXICO

Face Value	Year	% Pure Silver	$10	$11	$12	$13	$14	$15	$16	$17	$18	$19	$20	$21	$22	$23	$24	$25	$26	$27	$28	$29	$30
			THE MARKET PRICE OF SILVER IN DOLLARS PER OUNCE																				
20¢	1920-43	.720	.77	.84	.92	1.00	1.07	1.15	1.23	1.30	1.38	1.46	1.55	1.61	1.69	1.77	1.84	1.92	2.00	2.08	2.15	2.23	2.31
25¢	1950-53	.300	.32	.35	.38	.41	.45	.48	.51	.54	.57	.61	.64	.67	.70	.73	.76	.80	.83	.86	.89	.92	.96
50¢	1919-45	.720	1.93	2.12	2.31	2.50	2.70	2.89	3.08	3.28	3.47	3.66	3.86	4.05	4.24	4.43	4.63	4.82	5.01	5.21	5.40	5.59	5.79
50¢	1950-51	.720	.64	.70	.77	.83	.90	.96	1.03	1.09	1.16	1.22	1.27	1.32	1.38	1.44	1.51	1.57	1.63	1.70	1.76	1.82	1.89
1 Peso	1920-45	.720	3.85	4.23	4.62	5.00	5.39	5.77	6.16	6.54	6.93	7.31	7.72	8.08	8.47	8.85	9.26	9.64	10.01	10.39	10.78	11.16	11.55
1 Peso	1947-49	.500	2.25	2.47	2.70	2.92	3.15	3.37	3.60	3.82	4.05	4.27	4.51	4.72	4.95	5.17	5.40	5.62	5.85	6.07	6.30	6.52	6.75
1 Peso	1950	.300	1.29	1.41	1.54	1.67	1.80	1.93	2.06	2.19	2.32	2.45	2.57	2.70	2.83	2.96	3.09	3.22	3.35	3.48	3.61	3.74	3.87
1 Peso	1957-67	.100	.51	.56	.61	.66	.71	.76	.81	.86	.91	.97	1.03	1.07	1.12	1.17	1.22	1.27	1.32	1.37	1.43	1.48	1.53
5 Peso	1947-48	.900	8.68	9.54	10.41	11.28	12.15	13.02	13.88	14.75	15.62	16.49	17.36	18.22	19.09	19.96	20.83	21.70	22.56	23.43	24.30	25.17	26.04
5 Peso	1950-54	.720	6.43	7.07	7.71	8.35	9.00	9.64	10.28	10.93	11.57	12.21	12.87	13.50	14.14	14.78	15.43	16.07	16.71	17.36	18.00	18.64	19.29
5 Peso	1955-59	.720	4.18	4.59	5.01	5.45	5.85	6.27	6.68	7.10	7.52	7.94	8.36	8.77	9.19	9.61	10.00	10.45	10.86	11.28	11.70	12.12	12.54
10 Peso	1955-60	.900	8.35	9.18	10.02	10.85	11.69	12.52	13.36	14.19	15.03	15.86	16.72	17.53	18.37	19.20	20.04	20.87	21.71	22.54	23.38	24.21	25.05
25 Peso	1968-72	.720	5.21	5.73	6.25	6.77	7.29	7.81	8.33	8.85	9.37	9.89	10.42	10.94	11.46	11.98	12.50	13.02	13.54	14.06	14.58	15.10	15.63
1 Onza	1949	.925	10.00	11.00	12.00	13.00	14.00	15.00	16.00	17.00	18.00	19.00	20.00	21.00	22.00	23.00	24.00	25.00	26.00	27.00	28.00	29.00	30.00

Face Value	Year	% Pure Silver	$31	$32	$33	$34	$35	$36	$37	$38	$39	$40	$41	$42	$43	$44	$45	$46	$47	$48	$49	$50
			THE MARKET PRICE OF SILVER IN DOLLARS PER OUNCE																			
20¢	1920-43	.720	2.38	2.46	2.54	2.61	2.69	2.77	2.84	2.92	3.00	3.08	3.16	3.24	3.32	3.39	3.47	3.55	3.63	3.70	3.78	3.86
25¢	1950-53	.300	.99	1.02	1.05	1.08	1.12	1.15	1.18	1.21	1.24	1.28	1.32	1.35	1.38	1.41	1.45	1.48	1.51	1.54	1.58	1.61
50¢	1919-45	.720	5.98	6.17	6.36	6.56	6.75	6.94	7.25	7.33	7.52	7.72	7.91	8.10	8.29	8.49	8.68	8.87	9.06	9.26	9.45	9.64
50¢	1950-51	.720	1.95	2.01	2.07	2.14	2.20	2.26	2.33	2.39	2.45	2.52	2.63	2.70	2.76	2.83	2.89	2.95	3.02	3.08	3.15	3.21
1 Peso	1920-45	.720	11.93	12.32	12.70	13.09	13.47	13.86	14.24	14.63	15.01	15.40	15.81	16.20	16.59	16.97	17.36	17.75	18.13	18.51	18.90	19.29
1 Peso	1947-49	.500	6.97	7.20	7.42	7.65	7.87	8.10	8.32	8.55	8.77	9.00	9.23	9.45	9.68	9.90	10.13	10.35	10.58	10.80	11.03	11.25
1 Peso	1950	.300	4.00	4.12	4.25	4.38	4.51	4.64	4.77	4.90	5.03	5.16	5.27	5.40	5.53	5.66	5.78	5.91	6.04	6.17	6.30	6.43
1 Peso	1957-67	.100	1.58	1.63	1.68	1.73	1.78	1.83	1.88	1.94	1.99	2.04	2.10	2.16	2.21	2.26	2.31	2.37	2.41	2.46	2.52	2.70
5 Peso	1947-48	.900	26.90	27.77	28.64	29.51	30.38	31.24	32.11	32.98	33.85	34.72	35.60	36.19	37.05	37.91	38.78	39.64	40.50	41.36	42.23	43.09
5 Peso	1950-54	.720	19.93	20.57	21.21	21.86	22.50	23.14	23.79	24.43	25.07	25.72	26.36	27.00	27.65	28.29	28.93	29.58	30.22	30.86	31.50	32.15
5 Peso	1955-59	.720	12.95	13.37	13.79	14.21	14.63	15.04	15.46	15.88	16.30	16.72	17.14	17.55	17.97	18.39	18.81	19.23	19.64	20.06	20.48	20.90
10 Peso	1955-60	.900	25.88	26.70	27.55	28.39	29.22	30.06	30.89	31.73	32.56	33.40	34.27	35.10	35.94	36.78	37.61	38.45	39.28	40.12	40.95	41.79
25 Peso	1968-72	.720	16.15	16.67	17.19	17.71	18.23	18.75	19.27	19.80	20.31	20.84	21.35	21.87	22.40	22.91	23.44	23.96	24.48	25.00	25.52	26.04
1 Onza	1949	.925	31.00	32.00	33.00	34.00	35.00	36.00	37.00	38.00	39.00	40.00	41.00	42.00	43.00	44.00	45.00	46.00	47.00	48.00	49.00	50.00

The Mexican 1 Onza is an investment or trade coin with no marked value. It contains one ounce of pure silver.

Weighing Precious Metals

There are four systems of weights in use in the Western World today. (In the Orient and the Middle East there are several more.) 1. The apothecaries weight system which has long been used in weighing drugs and medicines. It is now being replaced by the metric system. 2. The avoirdupois system is used to weigh almost everything except precious metals, gemstones, medicines and drugs. 3. The troy system which is used only to weigh precious metals. It is believed the name came from a weight used at the fair of Troyes in France and it came to be known as the troy pound. There is a reference to the use of the troy pound as early as 1390 A.D. At this early date it was used to weigh silver and gold and to check the weight of money. It has become the standard system of weights used for precious metals in the United States and Great Britain.

The ounce and the pound are different weights in the apothecaries, the avoirdupois and the troy systems but the grain, the basic weight in all three, is the same. 4. The metric system developed in France, has become the system of weights and measures for the world scientific community. It is the most convenient system to use and one which may be used in comparing the weights of the other systems.

Although the world price for gold and silver is quoted for the troy ounce, many countries that use the metric system use the gram for weighing precious metals instead of the pennyweight. Because of this, there are included in this book two separate sets of tables for calculating the price of gold in either grams or pennyweights each in eight different karat values.

Pennies as a Substitute for Pennyweights

A new U.S. one cent copper penny weighs approximately 2 pennyweights, or 48 grains, and 10 pennies will weigh about 1 troy ounce or 480 grains. However each penny is not exactly the same weight because a 2 grain plus or minus (tolerance) is allowed when they are manufactured. This should average out when a number of new pennies are used to weigh an ounce or more but, because of this variance in weight, there could be a difference of almost 5 percent. Therefore when used for weighing, pennies should not be considered as accurate weights but only for close approximation.

SPECIFIC GRAVITIES

Metal	Specific gravity
Platinum	21.45
Platinum, 90%, iridium, 10%	21.5
Iridium	22.4
Platinum, 80%, palladium, 20%	18.3
Palladium	12.0
Rhodium	12.5
Gold (24 k.)	19.3
18 k. green	approx. 15.8
18 k. yellow	approx. 15.1
14 k. green	approx. 13.3
14 k. yellow	approx. 11.7
Silver (pure)	10.5
Sterling silver	10.3-10.4
Nickel	8.86
Monel metal (nickel-copper)	8.8
Stainless steel	7.7-7.9
Brass	approx. 8.4-8.8
Copper	8.9
Iron and steel	7.8-7.9
Lead	11.3
Zinc	7.14
German silver (nickel silver) (copper, zinc, nickel)	approx. 8.7
Aluminum bronze (dirigold) (copper, aluminum)	7.5-8.2
Bronze (copper, tin)	approx. 8.8-8.9
Tin	7.3

The accompaning table gives the specific gravities of the more common metals and alloys that are possible to come in contact with when checking the specific gravity of articles of jewelry. With absolutely pure metal the reading of their exact specific gravities is probable. The accuracy of the specific gravity reading will depend upon the accuracy of the scale or ballance and the care with which the exact weight readings are taken. With the alloys of gold and silver the specific gravity readings will usually be a close approximation as the table indicates. Even with the strict alloying of the metal used to make silver coins a variation is usually found. Gold coins are usually very close to the expected specific gravity. The specific gravity test is considered a very accurate way to identify metals and alloys.

You do not need to be a mathematician to find the value of the gold in an article of jewelry when the price per pennyweight charts on the following pages are used.

Tables For Calculating The Value Of Gold In Pennyweights

On the following pages the eight different gold karat tables are calculated to give the price in dollars per pennyweight for an article of gold when the price of gold is from 100 to 1000 dollars per ounce. This is the value of the pure gold in each pennyweight.

The gold price is in five dollar increments, or steps, and a price per pennyweight falling between the listed gold prices can easily be estimated.

To find the value of the gold in an article of jewelry three facts must be known.

The karat or the fineness of the gold in the article.

The weight of the article in pennyweights. (For gram weights use the gram tables.)

The market price of gold.

Use the gold karat table that corresponds with the karat of the gold article. Take the price per pennyweight next to the gold price. Multiply that figure by the weight in pennyweights of the article. The answer will give the value in dollars of the pure gold in an article.

For example, a 14 karat gold ring weighs 10 pennyweights and the market price of gold is $500; a simple multiplication will give the value. On the 14 karat gold pennyweight table the figure opposite the $500 gold price is 14.75. This price per pennyweight figure of 14.75 multiplied by the weight of the ring which is 10 pennyweights gives the answer of $147.50 as the value of the pure gold in the ring.

The value of the pure gold content in an article being sold for scrap metal cannot be considered the fair market value for that article. This gold value will necessarily be reduced, sometimes 30 or 40 percent, to account for refining, shipping, handling and a profit for the buyer or dealer. The percentage or the price paid for an article of gold jewelry is entirely up to the discretion of the buyer.

(There are also in this book eight gold karat tables calculated in grams to use when gold is weighed in the metric system.)

PRICE PER PENNYWEIGHT (dwt.) FOR 10K GOLD

Gold Price	Price Per DWT
100.00	2.08
105.00	2.18
110.00	2.29
115.00	2.39
120.00	2.50
125.00	2.60
130.00	2.70
135.00	2.81
140.00	2.91
145.00	3.02
150.00	3.12
155.00	3.23
160.00	3.33
165.00	3.44
170.00	3.54
175.00	3.64
180.00	3.75
185.00	3.85
190.00	3.95
195.00	4.06
200.00	4.16
205.00	4.27
210.00	4.37
215.00	4.47
220.00	4.58
225.00	4.69
230.00	4.79
235.00	4.89
240.00	5.00
245.00	5.10
250.00	5.20
255.00	5.31
260.00	5.41
265.00	5.52
270.00	5.62
275.00	5.72
280.00	5.83

Gold Price	Price Per DWT
285.00	5.93
290.00	6.04
295.00	6.14
300.00	6.25
305.00	6.35
310.00	6.45
315.00	6.56
320.00	6.66
325.00	6.77
330.00	6.87
335.00	6.98
340.00	7.08
345.00	7.18
350.00	7.29
355.00	7.39
360.00	7.50
365.00	7.60
370.00	7.70
375.00	7.81
380.00	7.91
385.00	8.02
390.00	8.12
395.00	8.23
400.00	8.33
405.00	8.43
410.00	8.54
415.00	8.64
420.00	8.75
425.00	8.85
430.00	8.95
435.00	9.06
440.00	9.16
445.00	9.27
450.00	9.37
455.00	9.48
460.00	9.58
465.00	9.68

Gold Price	Price Per DWT
470.00	9.79
475.00	9.89
480.00	10.00
485.00	10.10
490.00	10.20
495.00	10.31
500.00	10.41
505.00	10.52
510.00	10.62
515.00	10.73
520.00	10.83
525.00	10.93
530.00	11.04
535.00	11.14
540.00	11.25
545.00	11.35
550.00	11.45
555.00	11.56
560.00	11.66
565.00	11.77
570.00	11.87
575.00	11.98
580.00	12.08
585.00	12.18
590.00	12.28
595.00	12.39
600.00	12.50
605.00	12.60
610.00	12.70
615.00	12.81
620.00	12.91
625.00	13.02
630.00	13.12
635.00	13.23
640.00	13.33
645.00	13.43
650.00	13.54

Gold Price	Price Per DWT
655.00	13.64
660.00	13.75
665.00	13.85
670.00	13.95
675.00	14.06
680.00	14.16
685.00	14.27
690.00	14.37
695.00	14.48
700.00	14.58
705.00	14.68
710.00	14.78
715.00	14.89
720.00	15.00
725.00	15.10
730.00	15.20
735.00	15.31
740.00	15.41
745.00	15.52
750.00	15.62
755.00	15.73
760.00	15.83
765.00	15.93
770.00	16.04
775.00	16.14
780.00	16.25
785.00	16.35
790.00	16.45
795.00	16.56
800.00	16.66
805.00	16.77
810.00	16.87
815.00	16.98
820.00	17.08
825.00	17.18
830.00	17.28
835.00	17.39

Gold Price	Price Per DWT
840.00	17.50
845.00	17.60
850.00	17.70
855.00	17.81
860.00	17.91
865.00	18.02
870.00	18.12
875.00	18.23
880.00	18.33
885.00	18.43
890.00	18.54
895.00	18.64
900.00	18.75
905.00	18.85
910.00	18.95
915.00	19.06
920.00	19.16
925.00	19.26
930.00	19.37
935.00	19.47
940.00	19.58
945.00	19.68
950.00	19.78
955.00	19.89
960.00	20.00
965.00	20.10
970.00	20.20
975.00	20.30
980.00	20.41
985.00	20.51
990.00	20.62
995.00	20.73
1000.00	20.83

PRICE PER PENNYWEIGHT (dwt.) FOR 12K GOLD

Gold Price	Price Per DWT	Gold Price	Price Per DWT	Gold Price	Price Per DWT	Gold Price	Price Per DWT	Gold Price	Price Per DWT
100.00	2.49	285.00	7.12	470.00	11.75	655.00	16.37	840.00	21.00
105.00	2.61	290.00	7.25	475.00	11.87	660.00	16.50	845.00	21.12
110.00	2.75	295.00	7.37	480.00	12.00	665.00	16.62	850.00	21.24
115.00	2.89	300.00	7.50	485.00	12.12	670.00	16.74	855.00	21.37
120.00	3.00	305.00	7.59	490.00	12.24	675.00	16.87	860.00	21.49
125.00	3.12	310.00	7.74	495.00	12.37	680.00	16.99	865.00	21.62
130.00	3.24	315.00	7.87	500.00	12.49	685.00	17.12	870.00	21.74
135.00	3.37	320.00	7.99	505.00	12.62	690.00	17.24	875.00	21.87
140.00	3.49	325.00	8.12	510.00	12.74	695.00	17.37	880.00	21.99
145.00	3.62	330.00	8.24	515.00	12.87	700.00	17.49	885.00	22.11
150.00	3.74	335.00	8.37	520.00	12.99	705.00	17.61	890.00	22.25
155.00	3.87	340.00	8.49	525.00	13.11	710.00	17.75	895.00	22.37
160.00	3.99	345.00	8.61	530.00	13.25	715.00	17.86	900.00	22.50
165.00	4.13	350.00	8.75	535.00	13.37	720.00	18.00	905.00	22.62
170.00	4.25	355.00	8.87	540.00	13.50	725.00	18.12	910.00	22.74
175.00	4.37	360.00	9.00	545.00	13.62	730.00	18.24	915.00	22.87
180.00	4.50	365.00	9.12	550.00	13.74	735.00	18.37	920.00	22.99
185.00	4.62	370.00	9.24	555.00	13.87	740.00	18.49	925.00	23.11
190.00	4.74	375.00	9.37	560.00	13.99	745.00	18.62	930.00	23.24
195.00	4.87	380.00	9.49	565.00	14.12	750.00	18.74	935.00	23.36
200.00	4.99	385.00	9.62	570.00	14.24	755.00	18.87	940.00	23.49
205.00	5.12	390.00	9.74	575.00	14.37	760.00	18.99	945.00	23.61
210.00	5.24	395.00	9.87	580.00	14.49	765.00	19.11	950.00	23.73
215.00	5.36	400.00	9.99	585.00	14.61	770.00	19.25	955.00	23.86
220.00	5.50	405.00	10.11	590.00	14.73	775.00	19.37	960.00	24.00
225.00	5.63	410.00	10.25	595.00	14.87	780.00	19.50	965.00	24.12
230.00	5.75	415.00	10.37	600.00	15.00	785.00	19.62	970.00	24.24
235.00	5.88	420.00	10.50	605.00	15.12	790.00	19.74	975.00	24.36
240.00	6.00	425.00	10.62	610.00	15.24	795.00	19.87	980.00	24.49
245.00	6.12	430.00	10.74	615.00	15.37	800.00	19.99	985.00	24.61
250.00	6.24	435.00	10.87	620.00	15.49	805.00	20.12	990.00	24.74
255.00	6.37	440.00	10.99	625.00	15.62	810.00	20.24	995.00	24.87
260.00	6.49	445.00	11.12	630.00	15.74	815.00	20.37	1000.00	24.99
265.00	6.62	450.00	11.24	635.00	15.87	820.00	20.49		
270.00	6.72	455.00	11.37	640.00	15.99	825.00	20.61		
275.00	6.86	460.00	11.49	645.00	16.11	830.00	20.73		
280.00	6.99	465.00	11.61	650.00	16.25	835.00	20.86		

PRICE PER PENNYWEIGHT (dwt.) FOR 14K GOLD

Gold Price	Price Per DWT	Gold Price	Price Per DWT	Gold Price	Price Per DWT	Gold Price	Price Per DWT	Gold Price	Price Per DWT
100.00	2.91	285.00	8.30	470.00	13.70	655.00	19.09	840.00	24.50
105.00	3.05	290.00	8.45	475.00	13.84	660.00	19.25	845.00	24.64
110.00	3.20	295.00	8.59	480.00	14.00	665.00	19.44	850.00	24.78
115.00	3.34	300.00	8.75	485.00	14.14	670.00	19.53	855.00	24.93
120.00	3.50	305.00	8.86	490.00	14.28	675.00	19.68	860.00	25.07
125.00	3.64	310.00	9.03	495.00	14.43	680.00	19.82	865.00	25.23
130.00	3.78	315.00	9.18	500.00	14.57	685.00	19.98	870.00	25.37
135.00	3.93	320.00	9.32	505.00	14.73	690.00	20.11	875.00	25.52
140.00	4.07	325.00	9.48	510.00	14.87	695.00	20.27	880.00	25.66
145.00	4.23	330.00	9.62	515.00	15.02	700.00	20.41	885.00	25.80
150.00	4.37	335.00	9.77	520.00	15.16	705.00	20.55	890.00	25.95
155.00	4.52	340.00	9.91	525.00	15.30	710.00	20.70	895.00	26.09
160.00	4.66	345.00	10.05	530.00	15.45	715.00	20.84	900.00	26.25
165.00	4.81	350.00	10.20	535.00	15.59	720.00	21.00	905.00	26.39
170.00	4.95	355.00	10.34	540.00	15.75	725.00	21.14	910.00	26.53
175.00	5.09	360.00	10.50	545.00	15.89	730.00	21.28	915.00	26.68
180.00	5.25	365.00	10.64	550.00	16.03	735.00	21.43	920.00	26.82
185.00	5.39	370.00	10.78	555.00	16.18	740.00	21.57	925.00	26.96
190.00	5.53	375.00	10.93	560.00	16.32	745.00	21.73	930.00	27.12
195.00	5.68	380.00	11.07	565.00	16.48	750.00	21.87	935.00	27.26
200.00	5.82	385.00	11.23	570.00	16.62	755.00	22.02	940.00	27.41
205.00	5.98	390.00	11.37	575.00	16.77	760.00	22.16	945.00	27.55
210.00	6.12	395.00	11.52	580.00	16.91	765.00	22.30	950.00	27.69
215.00	6.26	400.00	11.66	585.00	17.05	770.00	22.45	955.00	27.84
220.00	6.41	405.00	11.80	590.00	17.19	775.00	22.59	960.00	28.00
225.00	6.56	410.00	11.95	595.00	17.34	780.00	22.75	965.00	28.14
230.00	6.70	415.00	12.09	600.00	17.50	785.00	22.87	970.00	28.28
235.00	6.84	420.00	12.25	605.00	17.64	790.00	23.03	975.00	28.42
240.00	7.00	425.00	12.39	610.00	17.78	795.00	23.18	980.00	28.57
245.00	7.14	430.00	12.53	615.00	17.93	800.00	23.32	985.00	28.71
250.00	7.28	435.00	12.68	620.00	18.07	805.00	23.48	990.00	28.86
255.00	7.43	440.00	12.82	625.00	18.23	810.00	23.62	995.00	29.02
260.00	7.57	445.00	12.98	630.00	18.37	815.00	23.77	1000.00	29.16
265.00	7.73	450.00	13.12	635.00	18.52	820.00	23.91		
270.00	7.87	455.00	13.27	640.00	18.66	825.00	24.05		
275.00	8.01	460.00	13.41	645.00	18.80	830.00	24.19		
280.00	8.16	465.00	13.55	650.00	18.95	835.00	24.34		

PRICE PER PENNYWEIGHT (dwt.) FOR 16K GOLD

Gold Price	Price Per DWT	Gold Price	Price Per DWT	Gold Price	Price Per DWT	Gold Price	Price Per DWT	Gold Price	Price Per DWT
100.00	3.33	285.00	9.49	470.00	15.66	655.00	21.82	840.00	28.00
105.00	3.48	290.00	9.66	475.00	15.82	660.00	22.00	845.00	28.16
110.00	3.66	295.00	9.82	480.00	16.00	665.00	22.16	850.00	28.32
115.00	3.82	300.00	10.00	485.00	16.16	670.00	22.32	855.00	28.49
120.00	4.00	305.00	10.13	490.00	16.32	675.00	22.49	860.00	28.65
125.00	4.16	310.00	10.32	495.00	16.49	680.00	22.65	865.00	28.83
130.00	4.32	315.00	10.49	500.00	16.65	685.00	22.83	870.00	28.99
135.00	4.49	320.00	10.65	505.00	16.83	690.00	22.99	875.00	29.17
140.00	4.65	325.00	10.83	510.00	16.99	695.00	23.17	880.00	29.33
145.00	4.83	330.00	10.99	515.00	17.17	700.00	23.33	885.00	29.49
150.00	4.99	335.00	11.17	520.00	17.33	705.00	23.49	890.00	29.66
155.00	5.17	340.00	11.33	525.00	17.49	710.00	23.66	895.00	29.82
160.00	5.33	345.00	11.49	530.00	17.66	715.00	23.82	900.00	30.00
165.00	5.50	350.00	11.66	535.00	17.82	720.00	24.00	905.00	30.16
170.00	5.66	355.00	11.82	540.00	18.00	725.00	24.16	910.00	30.32
175.00	5.82	360.00	12.00	545.00	18.16	730.00	24.32	915.00	30.49
180.00	6.00	365.00	12.16	550.00	18.32	735.00	24.49	920.00	30.65
185.00	6.16	370.00	12.32	555.00	18.49	740.00	24.65	925.00	30.81
190.00	6.32	375.00	12.49	560.00	18.65	745.00	24.83	930.00	30.99
195.00	6.49	380.00	12.66	565.00	18.83	750.00	24.99	935.00	31.15
200.00	6.65	385.00	12.83	570.00	18.99	755.00	25.17	940.00	31.33
205.00	6.83	390.00	12.99	575.00	19.17	760.00	25.33	945.00	31.49
210.00	6.99	395.00	13.17	580.00	19.33	765.00	25.49	950.00	31.65
215.00	7.15	400.00	13.33	585.00	19.49	770.00	25.66	955.00	31.82
220.00	7.33	405.00	13.49	590.00	19.65	775.00	25.82	960.00	32.00
225.00	7.50	410.00	13.66	595.00	19.82	780.00	26.00	965.00	32.16
230.00	7.66	415.00	13.82	600.00	20.00	785.00	26.16	970.00	32.32
235.00	7.82	420.00	14.00	605.00	20.16	790.00	26.32	975.00	32.48
240.00	8.00	425.00	14.16	610.00	20.32	795.00	26.49	980.00	32.65
245.00	8.16	430.00	14.32	615.00	20.49	800.00	26.65	985.00	32.81
250.00	8.32	435.00	14.49	620.00	20.65	805.00	26.83	990.00	32.99
255.00	8.49	440.00	14.65	625.00	20.83	810.00	26.99	995.00	33.17
260.00	8.65	445.00	14.83	630.00	20.99	815.00	27.16	1000.00	33.33
265.00	8.83	450.00	14.99	635.00	21.17	820.00	27.33		
270.00	8.99	455.00	15.16	640.00	21.33	825.00	27.49		
275.00	9.15	460.00	15.33	645.00	21.49	830.00	27.65		
280.00	9.33	465.00	15.49	650.00	21.66	835.00	27.82		

PRICE PER PENNYWEIGHT (dwt.) FOR 18K GOLD

Gold Price	Price Per DWT	Gold Price	Price Per DWT	Gold Price	Price Per DWT	Gold Price	Price Per DWT	Gold Price	Price Per DWT
100.00	3.74	285.00	10.67	470.00	17.62	655.00	24.55	840.00	31.50
105.00	3.92	290.00	10.87	475.00	17.80	660.00	24.75	845.00	31.68
110.00	4.12	295.00	11.05	480.00	18.00	665.00	24.93	850.00	31.86
115.00	4.30	300.00	11.25	485.00	18.18	670.00	25.11	855.00	32.06
120.00	4.50	305.00	11.39	490.00	18.36	675.00	25.31	860.00	32.24
125.00	4.68	310.00	11.61	495.00	18.59	680.00	25.49	865.00	32.43
130.00	4.86	315.00	11.80	500.00	18.72	685.00	25.68	870.00	32.61
135.00	5.06	320.00	11.99	505.00	18.93	690.00	25.86	875.00	32.81
140.00	5.24	325.00	12.18	510.00	19.11	695.00	26.06	880.00	32.99
145.00	5.43	330.00	12.36	515.00	19.31	700.00	26.24	885.00	33.17
150.00	5.61	335.00	12.56	520.00	19.49	705.00	26.42	890.00	33.37
155.00	5.81	340.00	12.74	525.00	19.67	710.00	26.62	895.00	33.55
160.00	5.99	345.00	12.92	530.00	19.87	715.00	26.80	900.00	33.75
165.00	6.19	350.00	13.12	535.00	20.05	720.00	27.00	905.00	33.93
170.00	6.37	355.00	13.30	540.00	20.25	725.00	27.18	910.00	34.11
175.00	6.55	360.00	13.50	545.00	20.43	730.00	27.36	915.00	34.31
180.00	6.75	365.00	13.68	550.00	20.61	735.00	27.59	920.00	34.49
185.00	6.93	370.00	13.86	555.00	20.81	740.00	27.74	925.00	34.67
190.00	7.11	375.00	14.06	560.00	20.99	745.00	27.93	930.00	34.86
195.00	7.31	380.00	14.24	565.00	21.18	750.00	28.11	935.00	35.04
200.00	7.49	385.00	14.43	570.00	21.36	755.00	28.31	940.00	35.24
205.00	7.68	390.00	14.61	575.00	21.56	760.00	28.49	945.00	35.42
210.00	7.86	395.00	14.81	580.00	21.74	765.00	28.67	950.00	35.60
215.00	8.04	400.00	14.99	585.00	21.92	770.00	28.87	955.00	35.80
220.00	8.24	405.00	15.17	590.00	22.10	775.00	29.05	960.00	36.00
225.00	8.44	410.00	15.37	595.00	22.30	780.00	29.25	965.00	36.18
230.00	8.62	415.00	15.55	600.00	22.50	785.00	29.43	970.00	36.36
235.00	8.80	420.00	15.75	605.00	22.68	790.00	29.61	975.00	36.54
240.00	9.00	425.00	15.93	610.00	22.86	795.00	29.80	980.00	36.74
245.00	9.18	430.00	16.11	615.00	23.06	800.00	29.99	985.00	36.92
250.00	9.36	435.00	16.31	620.00	23.24	805.00	30.18	990.00	37.11
255.00	9.56	440.00	16.49	625.00	23.43	810.00	30.36	995.00	37.31
260.00	9.74	445.00	16.68	630.00	23.61	815.00	30.56	1000.00	37.49
265.00	9.93	450.00	16.86	635.00	23.81	820.00	30.74		
270.00	10.12	455.00	17.06	640.00	23.99	825.00	30.92		
275.00	10.29	460.00	17.24	645.00	24.17	830.00	31.10		
280.00	10.49	465.00	17.42	650.00	24.37	835.00	31.30		

PRICE PER PENNYWEIGHT (dwt.) FOR 20K GOLD

Gold Price	Price Per DWT
100.00	4.16
105.00	4.36
110.00	4.58
115.00	4.78
120.00	5.00
125.00	5.20
130.00	5.40
135.00	5.62
140.00	5.82
145.00	6.04
150.00	6.24
155.00	6.46
160.00	6.66
165.00	6.88
170.00	7.08
175.00	7.28
180.00	7.50
185.00	7.70
190.00	7.90
195.00	8.12
200.00	8.32
205.00	8.54
210.00	8.74
215.00	8.94
220.00	9.16
225.00	9.38
230.00	9.58
235.00	9.78
240.00	10.00
245.00	10.20
250.00	10.40
255.00	10.62
260.00	10.82
265.00	11.04
270.00	11.24
275.00	11.44
280.00	11.66

Gold Price	Price Per DWT
285.00	11.86
290.00	12.08
295.00	12.28
300.00	12.50
305.00	12.66
310.00	12.90
315.00	13.12
320.00	13.32
325.00	13.54
330.00	13.74
335.00	13.96
340.00	14.16
345.00	14.36
350.00	14.58
355.00	14.78
360.00	15.00
365.00	15.20
370.00	15.40
375.00	15.62
380.00	15.82
385.00	16.04
390.00	16.24
395.00	16.46
400.00	16.66
405.00	16.86
410.00	17.08
415.00	17.28
420.00	17.50
425.00	17.70
430.00	17.90
435.00	18.12
440.00	18.32
445.00	18.54
450.00	18.74
455.00	18.96
460.00	19.16
465.00	19.36

Gold Price	Price Per DWT
470.00	19.58
475.00	19.78
480.00	20.00
485.00	20.20
490.00	20.40
495.00	20.62
500.00	20.82
505.00	21.04
510.00	21.24
515.00	21.46
520.00	21.66
525.00	21.86
530.00	22.08
535.00	22.28
540.00	22.50
545.00	22.70
550.00	22.90
555.00	23.12
560.00	23.32
565.00	23.54
570.00	23.74
575.00	23.96
580.00	24.16
585.00	24.36
590.00	24.56
595.00	24.78
600.00	25.00
605.00	25.20
610.00	25.40
615.00	25.62
620.00	25.82
625.00	26.04
630.00	26.24
635.00	26.46
640.00	26.66
645.00	26.86
650.00	27.08

Gold Price	Price Per DWT
655.00	27.28
660.00	27.50
665.00	27.70
670.00	27.90
675.00	28.12
680.00	28.32
685.00	28.54
690.00	28.74
695.00	28.96
700.00	29.16
705.00	29.36
710.00	29.58
715.00	29.78
720.00	30.00
725.00	30.20
730.00	30.40
735.00	30.62
740.00	30.82
745.00	31.04
750.00	31.24
755.00	31.46
760.00	31.66
765.00	31.86
770.00	32.08
775.00	32.28
780.00	32.50
785.00	32.70
790.00	32.90
795.00	33.12
800.00	33.32
805.00	33.54
810.00	33.74
815.00	33.96
820.00	34.16
825.00	34.36
830.00	34.56
835.00	34.78

Gold Price	Price Per DWT
840.00	35.00
845.00	35.20
850.00	35.40
855.00	35.62
860.00	35.82
865.00	36.04
870.00	36.24
875.00	36.46
880.00	36.66
885.00	36.86
890.00	37.08
895.00	37.28
900.00	37.50
905.00	37.70
910.00	37.90
915.00	38.12
920.00	38.32
925.00	38.52
930.00	38.74
935.00	38.94
940.00	39.16
945.00	39.36
950.00	39.56
955.00	39.78
960.00	40.00
965.00	40.20
970.00	40.40
975.00	40.60
980.00	40.82
985.00	41.02
990.00	41.24
995.00	41.46
1000.00	41.66

PRICE PER PENNYWEIGHT (dwt.) FOR 22K GOLD

Gold Price	Price Per DWT
100.00	4.57
105.00	4.79
110.00	5.04
115.00	5.26
120.00	5.50
125.00	5.72
130.00	5.94
135.00	6.18
140.00	6.40
145.00	6.64
150.00	6.86
155.00	7.10
160.00	7.32
165.00	7.57
170.00	7.79
175.00	8.00
180.00	8.25
185.00	8.47
190.00	8.69
195.00	8.93
200.00	9.15
205.00	9.39
210.00	9.61
215.00	9.83
220.00	10.07
225.00	10.32
230.00	10.54
235.00	10.76
240.00	11.00
245.00	11.22
250.00	11.44
255.00	11.68
260.00	11.90
265.00	12.14
270.00	12.36
275.00	12.58
280.00	12.82

Gold Price	Price Per DWT
285.00	13.04
290.00	13.29
295.00	13.51
300.00	13.75
305.00	13.92
310.00	14.19
315.00	14.43
320.00	14.65
325.00	14.89
330.00	15.11
335.00	15.35
340.00	15.57
345.00	15.79
350.00	16.04
355.00	16.26
360.00	16.50
365.00	16.72
370.00	16.94
375.00	17.18
380.00	17.40
385.00	17.64
390.00	17.86
395.00	18.10
400.00	18.32
405.00	18.54
410.00	18.79
415.00	19.00
420.00	19.25
425.00	19.47
430.00	19.69
435.00	19.93
440.00	20.15
445.00	20.39
450.00	20.61
455.00	20.85
460.00	21.07
465.00	21.29

Gold Price	Price Per DWT
470.00	21.54
475.00	21.76
480.00	22.00
485.00	22.22
490.00	22.44
495.00	22.68
500.00	22.90
505.00	23.14
510.00	23.36
515.00	23.60
520.00	23.82
525.00	24.04
530.00	24.29
535.00	24.51
540.00	24.75
545.00	24.97
550.00	25.19
555.00	25.43
560.00	25.65
565.00	25.89
570.00	26.11
575.00	26.35
580.00	26.57
585.00	26.79
590.00	27.01
595.00	27.26
600.00	27.50
605.00	27.72
610.00	27.94
615.00	28.18
620.00	28.40
625.00	28.64
630.00	28.86
635.00	29.10
640.00	29.32
645.00	29.54
650.00	29.78

Gold Price	Price Per DWT
655.00	30.00
660.00	30.25
665.00	30.47
670.00	30.69
675.00	30.93
680.00	31.15
685.00	31.39
690.00	31.61
695.00	31.85
700.00	32.07
705.00	32.29
710.00	32.54
715.00	32.76
720.00	33.00
725.00	33.22
730.00	33.44
735.00	33.68
740.00	33.90
745.00	34.14
750.00	34.36
755.00	34.60
760.00	34.82
765.00	35.04
770.00	35.28
775.00	35.50
780.00	35.75
785.00	35.97
790.00	36.19
795.00	36.43
800.00	36.65
805.00	36.89
810.00	37.11
815.00	37.35
820.00	37.57
825.00	37.79
830.00	38.01
835.00	38.26

Gold Price	Price Per DWT
840.00	38.50
845.00	38.72
850.00	38.94
855.00	39.18
860.00	39.40
865.00	39.64
870.00	39.86
875.00	40.10
880.00	40.32
885.00	40.54
890.00	40.79
895.00	41.00
900.00	41.25
905.00	41.47
910.00	41.69
915.00	41.93
920.00	42.15
925.00	42.37
930.00	42.61
935.00	42.83
940.00	43.07
945.00	43.29
950.00	43.51
955.00	43.75
960.00	44.00
965.00	44.22
970.00	44.44
975.00	44.66
980.00	44.90
985.00	45.12
990.00	45.36
995.00	45.60
1000.00	45.82

PRICE PER PENNYWEIGHT (dwt.) FOR 24K GOLD

Gold Price	Price Per DWT	Gold Price	Price Per DWT	Gold Price	Price Per DWT	Gold Price	Price Per DWT	Gold Price	Price Per DWT
100.00	4.99	285.00	14.23	470.00	23.49	655.00	32.73	840.00	42.00
105.00	5.23	290.00	14.49	475.00	23.73	660.00	33.00	845.00	42.24
110.00	5.49	295.00	14.73	480.00	24.00	665.00	33.24	850.00	42.48
115.00	5.73	300.00	15.00	485.00	24.24	670.00	33.48	855.00	42.74
120.00	6.00	305.00	15.19	490.00	24.48	675.00	33.74	860.00	42.98
125.00	6.24	310.00	15.48	495.00	24.74	680.00	33.98	865.00	43.25
130.00	6.48	315.00	15.74	500.00	24.98	685.00	34.25	870.00	43.49
135.00	6.74	320.00	15.98	505.00	25.25	690.00	34.49	875.00	43.75
140.00	6.98	325.00	16.25	510.00	25.49	695.00	34.75	880.00	43.99
145.00	7.25	330.00	16.49	515.00	25.75	700.00	34.99	885.00	44.23
150.00	7.49	335.00	16.75	520.00	25.99	705.00	35.23	890.00	44.49
155.00	7.75	340.00	16.99	525.00	26.23	710.00	35.49	895.00	44.73
160.00	7.99	345.00	17.23	530.00	26.49	715.00	35.73	900.00	45.00
165.00	8.25	350.00	17.49	535.00	26.73	720.00	36.00	905.00	45.24
170.00	8.49	355.00	17.73	540.00	27.00	725.00	36.24	910.00	45.48
175.00	8.73	360.00	18.00	545.00	27.24	730.00	36.48	915.00	45.74
180.00	9.00	365.00	18.24	550.00	27.48	735.00	36.74	920.00	45.98
185.00	9.24	370.00	18.48	555.00	27.74	740.00	36.98	925.00	46.22
190.00	9.48	375.00	18.74	560.00	27.98	745.00	37.25	930.00	46.49
195.00	9.74	380.00	18.98	565.00	28.25	750.00	37.49	935.00	46.73
200.00	9.98	385.00	19.25	570.00	28.49	755.00	37.75	940.00	46.99
205.00	10.25	390.00	19.49	575.00	28.75	760.00	37.99	945.00	47.23
210.00	10.49	395.00	19.75	580.00	28.99	765.00	38.23	950.00	47.47
215.00	10.73	400.00	19.99	585.00	29.23	770.00	38.49	955.00	47.73
220.00	10.99	405.00	20.23	590.00	29.47	775.00	38.73	960.00	48.00
225.00	11.25	410.00	20.49	595.00	29.73	780.00	39.00	965.00	48.24
230.00	11.49	415.00	20.73	600.00	30.00	785.00	39.24	970.00	48.48
235.00	11.73	420.00	21.00	605.00	30.24	790.00	39.48	975.00	48.72
240.00	12.00	425.00	21.24	610.00	30.48	795.00	39.74	980.00	48.98
245.00	12.24	430.00	21.48	615.00	30.74	800.00	39.98	985.00	49.22
250.00	12.48	435.00	21.74	620.00	30.98	805.00	40.25	990.00	49.49
255.00	12.74	440.00	21.98	625.00	31.25	810.00	40.49	995.00	49.75
260.00	12.98	445.00	22.25	630.00	31.49	815.00	40.75	1000.00	49.99
265.00	13.25	450.00	22.49	635.00	31.75	820.00	40.99		
270.00	13.49	455.00	22.75	640.00	31.99	825.00	41.23		
275.00	13.73	460.00	22.99	645.00	32.23	830.00	41.47		
280.00	13.99	465.00	23.23	650.00	32.49	835.00	41.73		

The use of the following price per gram tables will take the mystery and frustration out of finding the value of the gold in an article of jewelry.

Tables For Calculating The Value Of Gold In Grams

On the following pages the eight different gold karat tables were calculated to give the price in dollars per gram for an article of gold when the price of gold is from 100 to 1000 dollars per ounce. This is the value of the pure gold in each gram. The gold price is in five dollar increments, or steps, and a price per gram falling between the listed gold prices can easily be estimated.

To find the value of the gold in an article of jewelry three facts must be known.

The karat or fineness of the gold in the article.

The weight of the article in grams. (For pennyweights use the pennyweight tables.)

The market price of gold.

Use the gold karat table that corresponds with the karat of the gold article. Take the price per gram next to the gold price. Multiply that figure by the weight in grams for the article. The answer will give the value in dollars of pure gold in an article.

For example, a 14 karat gold ring weighs 10 grams and the market price of gold is $500; a simple multiplication will give the value. On the 14 karat gram table the figure opposite the $500 gold price is 9.41. This price per gram figure of 9.41 multiplied by the weight of the ring which is 10 grams, gives the answer of $94.10 as the value of the pure gold in the ring.

The value of the pure gold content in an article being sold for scrap metal cannot be considered the fair market value for that article. This gold value will necessarily be reduced, sometimes 30 or 40 percent to account for refining, shipping, handling and a profit for the buyer or dealer.

Also included in this book are eight gold karat tables calculated in pennyweights to use when gold is weighed in the troy system.

PRICE PER GRAM FOR 10K GOLD

Gold Price	Price Per Gram	Gold Price	Price Per Gram	Gold Price	Price Per Gram	Gold Price	Price Per Gram	Gold Price	Price Per Gram
100.00	1.34	285.00	3.83	470.00	6.31	655.00	8.80	840.00	11.29
105.00	1.41	290.00	3.89	475.00	6.38	660.00	8.87	845.00	11.35
110.00	1.48	295.00	3.96	480.00	6.45	665.00	8.93	850.00	11.42
115.00	1.54	300.00	4.03	485.00	6.51	670.00	9.00	855.00	11.49
120.00	1.61	305.00	4.10	490.00	6.58	675.00	9.07	860.00	11.56
125.00	1.68	310.00	4.16	495.00	6.65	680.00	9.14	865.00	11.62
130.00	1.75	315.00	4.23	500.00	6.72	685.00	9.20	870.00	11.69
135.00	1.81	320.00	4.30	505.00	6.79	690.00	9.27	875.00	11.76
140.00	1.88	325.00	4.37	510.00	6.85	695.00	9.34	880.00	11.83
145.00	1.95	330.00	4.43	515.00	6.92	700.00	9.41	885.00	11.89
150.00	2.01	335.00	4.50	520.00	6.99	705.00	9.47	890.00	11.96
155.00	2.08	340.00	4.57	525.00	7.05	710.00	9.54	895.00	12.02
160.00	2.15	345.00	4.64	530.00	7.12	715.00	9.61	900.00	12.09
165.00	2.21	350.00	4.70	535.00	7.19	720.00	9.68	905.00	12.16
170.00	2.28	355.00	4.77	540.00	7.25	725.00	9.74	910.00	12.23
175.00	2.35	360.00	4.83	545.00	7.32	730.00	9.81	915.00	12.29
180.00	2.42	365.00	4.90	550.00	7.39	735.00	9.88	920.00	12.36
185.00	2.48	370.00	4.97	555.00	7.46	740.00	9.94	925.00	12.43
190.00	2.55	375.00	5.04	560.00	7.52	745.00	10.01	930.00	12.50
195.00	2.62	380.00	5.10	565.00	7.59	750.00	10.08	935.00	12.56
200.00	2.69	385.00	5.17	570.00	7.66	755.00	10.14	940.00	12.63
205.00	2.75	390.00	5.24	575.00	7.73	760.00	10.21	945.00	12.70
210.00	2.82	395.00	5.31	580.00	7.79	765.00	10.28	950.00	12.76
215.00	2.89	400.00	5.37	585.00	7.86	770.00	10.34	955.00	12.83
220.00	2.95	405.00	5.44	590.00	7.93	775.00	10.41	960.00	12.90
225.00	3.02	410.00	5.51	595.00	7.99	780.00	10.48	965.00	12.97
230.00	3.09	415.00	5.58	600.00	8.06	785.00	10.55	970.00	13.03
235.00	3.16	420.00	5.64	605.00	8.13	790.00	10.61	975.00	13.10
240.00	3.22	425.00	5.71	610.00	8.20	795.00	10.68	980.00	13.17
245.00	3.29	430.00	5.78	615.00	8.26	800.00	10.75	985.00	13.24
250.00	3.36	435.00	5.84	620.00	8.33	805.00	10.81	990.00	13.30
255.00	3.43	440.00	5.91	625.00	8.40	810.00	10.88	995.00	13.37
260.00	3.49	445.00	5.98	630.00	8.47	815.00	10.95	1000.00	13.44
265.00	3.56	450.00	6.05	635.00	8.53	820.00	11.02		
270.00	3.63	455.00	6.11	640.00	8.60	825.00	11.08		
275.00	3.69	460.00	6.18	645.00	8.67	830.00	11.15		
280.00	3.76	465.00	6.25	650.00	8.73	835.00	11.22		

PRICE PER GRAM FOR 12K GOLD

Gold Price	Price Per Gram
100.00	1.61
105.00	1.69
110.00	1.77
115.00	1.85
120.00	1.93
125.00	2.01
130.00	2.10
135.00	2.17
140.00	2.25
145.00	2.34
150.00	2.41
155.00	2.49
160.00	2.58
165.00	2.65
170.00	2.73
175.00	2.82
180.00	2.90
185.00	2.97
190.00	3.06
195.00	3.14
200.00	3.23
205.00	3.30
210.00	3.38
215.00	3.47
220.00	3.54
225.00	3.62
230.00	3.71
235.00	3.79
240.00	3.86
245.00	3.95
250.00	4.03
255.00	4.11
260.00	4.19
265.00	4.27
270.00	4.35
275.00	4.43
280.00	4.51

Gold Price	Price Per Gram
285.00	4.59
290.00	4.67
295.00	4.75
300.00	4.83
305.00	4.92
310.00	4.99
315.00	5.07
320.00	5.16
325.00	5.24
330.00	5.31
335.00	5.40
340.00	5.48
345.00	5.57
350.00	5.64
355.00	5.72
360.00	5.79
365.00	5.88
370.00	5.96
375.00	6.05
380.00	6.12
385.00	6.20
390.00	6.29
395.00	6.37
400.00	6.44
405.00	6.53
410.00	6.61
415.00	6.69
420.00	6.77
425.00	6.85
430.00	6.93
435.00	7.01
440.00	7.09
445.00	7.17
450.00	7.26
455.00	7.33
460.00	7.41
465.00	7.50

Gold Price	Price Per Gram
470.00	7.57
475.00	7.65
480.00	7.74
485.00	7.81
490.00	7.89
495.00	7.98
500.00	8.06
505.00	8.15
510.00	8.22
515.00	8.30
520.00	8.39
525.00	8.46
530.00	8.54
535.00	8.63
540.00	8.70
545.00	8.78
550.00	8.87
555.00	8.95
560.00	9.02
565.00	9.11
570.00	9.19
575.00	9.27
580.00	9.35
585.00	9.43
590.00	9.51
595.00	9.59
600.00	9.67
605.00	9.75
610.00	9.84
615.00	9.91
620.00	9.99
625.00	10.08
630.00	10.16
635.00	10.23
640.00	10.32
645.00	10.40
650.00	10.47

Gold Price	Price Per Gram
655.00	10.56
660.00	10.64
665.00	10.71
670.00	10.80
675.00	10.88
680.00	10.97
685.00	11.04
690.00	11.12
695.00	11.20
700.00	11.29
705.00	11.36
710.00	11.45
715.00	11.53
720.00	11.61
725.00	11.69
730.00	11.77
735.00	11.85
740.00	11.93
745.00	12.01
750.00	12.09
755.00	12.17
760.00	12.25
765.00	12.33
770.00	12.41
775.00	12.49
780.00	12.57
785.00	12.66
790.00	12.73
795.00	12.81
800.00	12.90
805.00	12.97
810.00	13.05
815.00	13.14
820.00	13.22
825.00	13.29
830.00	13.38
835.00	13.46

Gold Price	Price Per Gram
840.00	13.55
845.00	13.62
850.00	13.70
855.00	13.79
860.00	13.87
865.00	13.94
870.00	14.03
875.00	14.11
880.00	14.19
885.00	14.27
890.00	14.35
895.00	14.42
900.00	14.51
905.00	14.59
910.00	14.67
915.00	14.75
920.00	14.83
925.00	14.91
930.00	15.00
935.00	15.07
940.00	15.15
945.00	15.24
950.00	15.31
955.00	15.39
960.00	15.48
965.00	15.56
970.00	15.63
975.00	15.72
980.00	15.80
985.00	15.89
990.00	15.96
995.00	16.04
1000.00	16.13

PRICE PER GRAM FOR 14K GOLD

Gold Price	Price Per Gram	Gold Price	Price Per Gram	Gold Price	Price Per Gram	Gold Price	Price Per Gram	Gold Price	Price Per Gram
100.00	1.87	285.00	5.36	470.00	8.83	655.00	12.32	840.00	15.80
105.00	1.97	290.00	5.44	475.00	8.93	660.00	12.42	845.00	15.89
110.00	2.07	295.00	5.54	480.00	9.03	665.00	12.50	850.00	15.99
115.00	2.15	300.00	5.64	485.00	9.11	670.00	12.60	855.00	16.08
120.00	2.25	305.00	5.74	490.00	9.21	675.00	12.70	860.00	16.18
125.00	2.35	310.00	5.82	495.00	9.31	680.00	12.79	865.00	16.27
130.00	2.45	315.00	5.92	500.00	9.41	685.00	12.88	870.00	16.36
135.00	2.53	320.00	6.02	505.00	9.50	690.00	12.98	875.00	16.46
140.00	2.63	325.00	6.12	510.00	9.59	695.00	13.07	880.00	16.56
145.00	2.73	330.00	6.20	515.00	9.69	700.00	13.17	885.00	16.64
150.00	2.81	335.00	6.30	520.00	9.78	705.00	13.26	890.00	16.74
155.00	2.91	340.00	6.40	525.00	9.87	710.00	13.35	895.00	16.83
160.00	3.01	345.00	6.49	530.00	9.97	715.00	13.45	900.00	16.92
165.00	3.09	350.00	6.58	535.00	10.06	720.00	13.55	905.00	17.02
170.00	3.19	355.00	6.68	540.00	10.15	725.00	13.63	910.00	17.12
175.00	3.29	360.00	6.76	545.00	10.25	730.00	13.73	915.00	17.20
180.00	3.39	365.00	6.86	550.00	10.34	735.00	13.83	920.00	17.30
185.00	3.47	370.00	6.96	555.00	10.44	740.00	13.91	925.00	17.40
190.00	3.57	375.00	7.05	560.00	10.53	745.00	14.01	930.00	17.50
195.00	3.67	380.00	7.14	565.00	10.62	750.00	14.11	935.00	17.58
200.00	3.76	385.00	7.24	570.00	10.72	755.00	14.19	940.00	17.68
205.00	3.85	390.00	7.33	575.00	10.82	760.00	14.29	945.00	17.78
210.00	3.95	395.00	7.43	580.00	10.90	765.00	14.39	950.00	17.86
215.00	4.04	400.00	7.52	585.00	11.00	770.00	14.47	955.00	17.96
220.00	4.13	405.00	7.61	590.00	11.10	775.00	14.57	960.00	18.06
225.00	4.23	410.00	7.71	595.00	11.18	780.00	14.67	965.00	18.16
230.00	4.32	415.00	7.81	600.00	11.28	785.00	14.77	970.00	18.24
235.00	4.42	420.00	7.89	605.00	11.38	790.00	14.85	975.00	18.34
240.00	4.51	425.00	7.99	610.00	11.48	795.00	14.95	980.00	18.44
245.00	4.60	430.00	8.09	615.00	11.56	800.00	15.05	985.00	18.53
250.00	4.70	435.00	8.17	620.00	11.66	805.00	15.13	990.00	18.62
255.00	4.80	440.00	8.27	625.00	11.76	810.00	15.23	995.00	18.72
260.00	4.88	445.00	8.37	630.00	11.86	815.00	15.33	1000.00	18.81
265.00	4.98	450.00	8.47	635.00	11.94	820.00	15.43		
270.00	5.08	455.00	8.55	640.00	12.04	825.00	15.51		
275.00	5.16	460.00	8.65	645.00	12.14	830.00	15.61		
280.00	5.26	465.00	8.75	650.00	12.22	835.00	15.71		

PRICE PER GRAM FOR 16K GOLD

Gold Price	Price Per Gram	Gold Price	Price Per Gram	Gold Price	Price Per Gram	Gold Price	Price Per Gram	Gold Price	Price Per Gram
100.00	2.14	285.00	6.13	470.00	10.09	655.00	14.08	840.00	18.06
105.00	2.25	290.00	6.22	475.00	10.21	660.00	14.19	845.00	18.16
110.00	2.37	295.00	6.33	480.00	10.32	665.00	14.29	850.00	18.27
115.00	2.46	300.00	6.45	485.00	10.41	670.00	14.40	855.00	18.38
120.00	2.57	305.00	6.56	490.00	10.53	675.00	14.51	860.00	18.49
125.00	2.69	310.00	6.65	495.00	10.64	680.00	14.62	865.00	18.59
130.00	2.80	315.00	6.77	500.00	10.75	685.00	14.72	870.00	18.70
135.00	2.89	320.00	6.88	505.00	10.86	690.00	14.83	875.00	18.81
140.00	3.01	325.00	6.99	510.00	10.96	695.00	14.94	880.00	18.93
145.00	3.12	330.00	7.09	515.00	11.07	700.00	15.05	885.00	19.02
150.00	3.21	335.00	7.20	520.00	11.18	705.00	15.15	890.00	19.13
155.00	3.33	340.00	7.31	525.00	11.28	710.00	15.26	895.00	19.23
160.00	3.44	345.00	7.42	530.00	11.39	715.00	15.37	900.00	19.34
165.00	3.53	350.00	7.52	535.00	11.50	720.00	15.49	905.00	19.45
170.00	3.65	355.00	7.63	540.00	11.60	725.00	15.58	910.00	19.57
175.00	3.76	360.00	7.73	545.00	11.71	730.00	15.69	915.00	19.66
180.00	3.87	365.00	7.84	550.00	11.82	735.00	15.81	920.00	19.77
185.00	3.97	370.00	7.95	555.00	11.93	740.00	15.90	925.00	19.89
190.00	4.08	375.00	8.06	560.00	12.03	745.00	16.01	930.00	20.00
195.00	4.19	380.00	8.16	565.00	12.14	750.00	16.13	935.00	20.09
200.00	4.30	385.00	8.27	570.00	12.25	755.00	16.22	940.00	20.21
205.00	4.40	390.00	8.38	575.00	12.37	760.00	16.33	945.00	20.32
210.00	4.51	395.00	8.49	580.00	12.46	765.00	16.45	950.00	20.41
215.00	4.62	400.00	8.59	585.00	12.57	770.00	16.54	955.00	20.53
220.00	4.72	405.00	8.70	590.00	12.69	775.00	16.65	960.00	20.64
225.00	4.83	410.00	8.81	595.00	12.78	780.00	16.77	965.00	20.75
230.00	4.94	415.00	8.93	600.00	12.89	785.00	16.88	970.00	20.85
235.00	5.05	420.00	9.02	605.00	13.01	790.00	16.97	975.00	20.96
240.00	5.15	425.00	9.13	610.00	13.12	795.00	17.09	980.00	21.07
245.00	5.26	430.00	9.25	615.00	13.21	800.00	17.20	985.00	21.18
250.00	5.37	435.00	9.34	620.00	13.33	805.00	17.29	990.00	21.28
255.00	5.49	440.00	9.45	625.00	13.44	810.00	17.41	995.00	21.39
260.00	5.58	445.00	9.57	630.00	13.55	815.00	17.52	1000.00	21.50
265.00	5.69	450.00	9.68	635.00	13.65	820.00	17.63		
270.00	5.81	455.00	9.77	640.00	13.76	825.00	17.73		
275.00	5.90	460.00	9.89	645.00	13.87	830.00	17.84		
280.00	6.01	465.00	10.00	650.00	13.97	835.00	17.95		

PRICE PER GRAM FOR 18K GOLD

Gold Price	Price Per Gram	Gold Price	Price Per Gram	Gold Price	Price Per Gram	Gold Price	Price Per Gram	Gold Price	Price Per Gram
100.00	2.41	285.00	6.89	470.00	11.36	655.00	15.84	840.00	20.32
105.00	2.54	290.00	7.00	475.00	11.48	660.00	15.96	845.00	20.43
110.00	2.66	295.00	7.13	480.00	11.61	665.00	16.07	850.00	20.55
115.00	2.77	300.00	7.25	485.00	11.72	670.00	16.20	855.00	20.68
120.00	2.90	305.00	7.38	490.00	11.84	675.00	16.32	860.00	20.81
125.00	3.02	310.00	7.49	495.00	11.97	680.00	16.45	865.00	20.91
130.00	3.15	315.00	7.61	500.00	12.09	685.00	16.56	870.00	21.04
135.00	3.26	320.00	7.74	505.00	12.22	690.00	16.68	875.00	21.17
140.00	3.38	325.00	7.86	510.00	12.33	695.00	16.81	880.00	21.29
145.00	3.51	330.00	7.97	515.00	12.45	700.00	16.94	885.00	21.40
150.00	3.62	335.00	8.10	520.00	12.58	705.00	17.04	890.00	21.53
155.00	3.74	340.00	8.22	525.00	12.69	710.00	17.17	895.00	21.63
160.00	3.87	345.00	8.35	530.00	12.81	715.00	17.30	900.00	21.76
165.00	3.98	350.00	8.46	535.00	12.94	720.00	17.42	905.00	21.89
170.00	4.10	355.00	8.58	540.00	13.05	725.00	17.53	910.00	22.01
175.00	4.23	360.00	8.69	545.00	13.17	730.00	17.66	915.00	22.12
180.00	4.35	365.00	8.82	550.00	13.30	735.00	17.78	920.00	22.25
185.00	4.46	370.00	8.94	555.00	13.43	740.00	17.89	925.00	22.37
190.00	4.59	375.00	9.07	560.00	13.53	745.00	18.02	930.00	22.50
195.00	4.71	380.00	9.18	565.00	13.66	750.00	18.14	935.00	22.61
200.00	4.84	385.00	9.30	570.00	13.79	755.00	18.25	940.00	22.73
205.00	4.95	390.00	9.43	575.00	13.91	760.00	18.38	945.00	22.86
210.00	5.07	395.00	9.56	580.00	14.02	765.00	18.50	950.00	22.97
215.00	5.20	400.00	9.66	585.00	14.15	770.00	18.61	955.00	23.09
220.00	5.31	405.00	9.79	590.00	14.27	775.00	18.74	960.00	23.22
225.00	5.43	410.00	9.92	595.00	14.38	780.00	18.86	965.00	23.34
230.00	5.56	415.00	10.04	600.00	14.51	785.00	18.99	970.00	23.45
235.00	5.69	420.00	10.15	605.00	14.63	790.00	19.10	975.00	23.58
240.00	5.79	425.00	10.28	610.00	14.76	795.00	19.22	980.00	23.70
245.00	5.92	430.00	10.40	615.00	14.87	800.00	19.35	985.00	23.83
250.00	6.05	435.00	10.51	620.00	14.99	805.00	19.46	990.00	23.94
255.00	6.17	440.00	10.64	625.00	15.12	810.00	19.58	995.00	24.06
260.00	6.28	445.00	10.76	630.00	15.24	815.00	19.71	1000.00	24.19
265.00	6.41	450.00	10.89	635.00	15.35	820.00	19.83		
270.00	6.53	455.00	11.00	640.00	15.48	825.00	19.94		
275.00	6.64	460.00	11.12	645.00	15.60	830.00	20.07		
280.00	6.77	465.00	11.25	650.00	15.71	835.00	20.19		

PRICE PER GRAM FOR 20K GOLD

Gold Price	Price Per Gram	Gold Price	Price Per Gram	Gold Price	Price Per Gram	Gold Price	Price Per Gram	Gold Price	Price Per Gram
100.00	2.68	285.00	7.66	470.00	12.62	655.00	17.60	840.00	22.58
105.00	2.82	290.00	7.78	475.00	12.76	660.00	17.74	845.00	22.70
110.00	2.96	295.00	7.92	480.00	12.90	665.00	17.86	850.00	22.84
115.00	3.08	300.00	8.06	485.00	13.02	670.00	18.00	855.00	22.98
120.00	3.22	305.00	8.20	490.00	13.16	675.00	18.14	860.00	23.12
125.00	3.36	310.00	8.32	495.00	13.30	680.00	18.28	865.00	23.24
130.00	3.50	315.00	8.46	500.00	13.44	685.00	18.40	870.00	23.38
135.00	3.62	320.00	8.60	505.00	13.58	690.00	18.54	875.00	23.52
140.00	3.76	325.00	8.74	510.00	13.70	695.00	18.68	880.00	23.66
145.00	3.90	330.00	8.86	515.00	13.84	700.00	18.82	885.00	23.78
150.00	4.02	335.00	9.00	520.00	13.98	705.00	18.94	890.00	23.92
155.00	4.16	340.00	9.14	525.00	14.10	710.00	19.08	895.00	24.04
160.00	4.30	345.00	9.28	530.00	14.24	715.00	19.22	900.00	24.18
165.00	4.42	350.00	9.40	535.00	14.38	720.00	19.36	905.00	24.32
170.00	4.56	355.00	9.54	540.00	14.50	725.00	19.48	910.00	24.46
175.00	4.70	360.00	9.66	545.00	14.64	730.00	19.62	915.00	24.58
180.00	4.84	365.00	9.80	550.00	14.78	735.00	19.76	920.00	24.72
185.00	4.96	370.00	9.94	555.00	14.92	740.00	19.88	925.00	24.86
190.00	5.10	375.00	10.08	560.00	15.04	745.00	20.02	930.00	25.00
195.00	5.24	380.00	10.20	565.00	15.18	750.00	20.16	935.00	25.12
200.00	5.38	385.00	10.34	570.00	15.32	755.00	20.28	940.00	25.26
205.00	5.50	390.00	10.48	575.00	15.46	760.00	20.42	945.00	25.40
210.00	5.64	395.00	10.62	580.00	15.58	765.00	20.56	950.00	25.52
215.00	5.78	400.00	10.74	585.00	15.72	770.00	20.68	955.00	25.66
220.00	5.90	405.00	10.88	590.00	15.86	775.00	20.82	960.00	25.80
225.00	6.04	410.00	11.02	595.00	15.98	780.00	20.96	965.00	25.94
230.00	6.18	415.00	11.16	600.00	16.12	785.00	21.10	970.00	26.06
235.00	6.32	420.00	11.28	605.00	16.26	790.00	21.22	975.00	26.20
240.00	6.44	425.00	11.42	610.00	16.40	795.00	21.36	980.00	26.34
245.00	6.58	430.00	11.56	615.00	16.52	800.00	21.50	985.00	26.48
250.00	6.72	435.00	11.68	620.00	16.66	805.00	21.62	990.00	26.60
255.00	6.86	440.00	11.82	625.00	16.80	810.00	21.76	995.00	26.74
260.00	6.98	445.00	11.96	630.00	16.94	815.00	21.90	1000.00	26.88
265.00	7.12	450.00	12.10	635.00	17.06	820.00	22.04		
270.00	7.26	455.00	12.22	640.00	17.20	825.00	22.16		
275.00	7.38	460.00	12.36	645.00	17.34	830.00	22.30		
280.00	7.52	465.00	12.50	650.00	17.46	835.00	22.44		

PRICE PER GRAM FOR 22K GOLD

Gold Price	Price Per Gram
100.00	2.95
105.00	3.10
110.00	3.25
115.00	3.39
120.00	3.54
125.00	3.69
130.00	3.85
135.00	3.98
140.00	4.13
145.00	4.29
150.00	4.42
155.00	4.57
160.00	4.73
165.00	4.86
170.00	5.01
175.00	5.17
180.00	5.32
185.00	5.45
190.00	5.61
195.00	5.76
200.00	5.92
205.00	6.05
210.00	6.20
215.00	6.36
220.00	6.49
225.00	6.64
230.00	6.80
235.00	6.95
240.00	7.08
245.00	7.24
250.00	7.39
255.00	7.54
260.00	7.68
265.00	7.83
270.00	7.98
275.00	8.12
280.00	8.27

Gold Price	Price Per Gram
285.00	8.42
290.00	8.56
295.00	8.71
300.00	8.86
305.00	9.02
310.00	9.15
315.00	9.30
320.00	9.46
325.00	9.61
330.00	9.74
335.00	9.90
340.00	10.05
345.00	10.21
350.00	10.34
355.00	10.49
360.00	10.62
365.00	10.78
370.00	10.93
375.00	11.09
380.00	11.22
385.00	11.37
390.00	11.53
395.00	11.68
400.00	11.81
405.00	11.96
410.00	12.12
415.00	12.27
420.00	12.41
425.00	12.56
430.00	12.71
435.00	12.85
440.00	13.00
445.00	13.15
450.00	13.31
455.00	13.44
460.00	13.59
465.00	13.75

Gold Price	Price Per Gram
470.00	13.88
475.00	14.03
480.00	14.19
485.00	14.32
490.00	14.47
495.00	14.63
500.00	14.78
505.00	14.94
510.00	15.07
515.00	15.22
520.00	15.38
525.00	15.51
530.00	15.66
535.00	15.82
540.00	15.95
545.00	16.10
550.00	16.26
555.00	16.41
560.00	16.54
565.00	16.70
570.00	16.85
575.00	17.00
580.00	17.14
585.00	17.29
590.00	17.44
595.00	17.58
600.00	17.73
605.00	17.88
610.00	18.04
615.00	18.17
620.00	18.32
625.00	18.48
630.00	18.63
635.00	18.76
640.00	18.92
645.00	19.07
650.00	19.20

Gold Price	Price Per Gram
655.00	19.36
660.00	19.51
665.00	19.64
670.00	19.80
675.00	19.95
680.00	20.11
685.00	20.24
690.00	20.39
695.00	20.55
700.00	20.70
705.00	20.83
710.00	20.99
715.00	21.14
720.00	21.29
725.00	21.43
730.00	21.58
735.00	21.73
740.00	21.87
745.00	22.02
750.00	22.17
755.00	22.31
760.00	22.46
765.00	22.61
770.00	22.75
775.00	22.90
780.00	23.05
785.00	23.21
790.00	23.34
795.00	23.49
800.00	23.65
805.00	23.78
810.00	23.93
815.00	24.09
820.00	24.24
825.00	24.37
830.00	24.53
835.00	24.68

Gold Price	Price Per Gram
840.00	24.84
845.00	24.97
850.00	25.12
855.00	25.28
860.00	25.43
865.00	25.56
870.00	25.72
875.00	25.87
880.00	26.02
885.00	26.16
890.00	26.31
895.00	26.44
900.00	26.60
905.00	26.75
910.00	26.90
915.00	27.04
920.00	27.19
925.00	27.36
930.00	27.50
935.00	27.63
940.00	27.78
945.00	27.94
950.00	28.07
955.00	28.22
960.00	28.38
965.00	28.53
970.00	28.66
975.00	28.82
980.00	28.97
985.00	29.13
990.00	29.26
995.00	29.41
1000.00	29.57

PRICE PER GRAM FOR 24K GOLD

Gold Price	Price Per Gram
100.00	3.21
105.00	3.38
110.00	3.55
115.00	3.69
120.00	3.86
125.00	4.03
130.00	4.20
135.00	4.34
140.00	4.51
145.00	4.68
150.00	4.82
155.00	4.99
160.00	5.16
165.00	5.30
170.00	5.47
175.00	5.64
180.00	5.81
185.00	5.95
190.00	6.12
195.00	6.29
200.00	6.45
205.00	6.60
210.00	6.79
215.00	6.93
220.00	7.08
225.00	7.25
230.00	7.41
235.00	7.58
240.00	7.73
245.00	7.89
250.00	8.06
255.00	8.23
260.00	8.37
265.00	8.54
270.00	8.71
275.00	8.85
280.00	9.02

Gold Price	Price Per Gram
285.00	9.19
290.00	9.33
295.00	9.50
300.00	9.67
305.00	9.84
310.00	9.98
315.00	10.15
320.00	10.32
325.00	10.49
330.00	10.63
335.00	10.80
340.00	10.97
345.00	11.13
350.00	11.28
355.00	11.45
360.00	11.59
365.00	11.76
370.00	11.93
375.00	12.09
380.00	12.24
385.00	12.41
390.00	12.57
395.00	12.74
400.00	12.88
405.00	13.05
410.00	13.22
415.00	13.39
420.00	13.53
425.00	13.70
430.00	13.87
435.00	14.01
440.00	14.18
445.00	14.35
450.00	14.52
455.00	14.66
460.00	14.83
465.00	15.00

Gold Price	Price Per Gram
470.00	15.14
475.00	15.31
480.00	15.48
485.00	15.62
490.00	15.79
495.00	15.96
500.00	16.13
505.00	16.29
510.00	16.44
515.00	16.60
520.00	16.77
525.00	16.92
530.00	17.08
535.00	17.25
540.00	17.40
545.00	17.57
550.00	17.73
555.00	17.90
560.00	18.05
565.00	18.21
570.00	18.38
575.00	18.55
580.00	18.69
585.00	18.86
590.00	19.03
595.00	19.17
600.00	19.34
605.00	19.51
610.00	19.68
615.00	19.82
620.00	19.99
625.00	20.16
630.00	20.32
635.00	20.47
640.00	20.64
645.00	20.80
650.00	20.95

Gold Price	Price Per Gram
655.00	21.12
660.00	21.29
665.00	21.43
670.00	21.60
675.00	21.77
680.00	21.93
685.00	22.08
690.00	22.25
695.00	22.41
700.00	22.58
705.00	22.73
710.00	22.89
715.00	23.06
720.00	23.23
725.00	23.37
730.00	23.54
735.00	23.71
740.00	23.85
745.00	24.02
750.00	24.19
755.00	24.33
760.00	24.50
765.00	24.67
770.00	24.81
775.00	24.98
780.00	25.15
785.00	25.32
790.00	25.46
795.00	25.63
800.00	25.80
805.00	25.94
810.00	26.11
815.00	26.28
820.00	26.45
825.00	26.59
830.00	26.76
835.00	26.93

Gold Price	Price Per Gram
840.00	27.09
845.00	27.24
850.00	27.41
855.00	27.57
860.00	27.74
865.00	27.89
870.00	28.05
875.00	28.22
880.00	28.39
885.00	28.53
890.00	28.70
895.00	28.85
900.00	29.01
905.00	29.18
910.00	29.35
915.00	29.49
920.00	29.66
925.00	29.83
930.00	30.00
935.00	30.14
940.00	30.31
945.00	30.48
950.00	30.62
955.00	30.79
960.00	30.96
965.00	31.13
970.00	31.27
975.00	31.44
980.00	31.61
985.00	31.77
990.00	31.92
995.00	32.08
1000.00	32.25

Safeguarding Your Gold And Silver

The safety deposit vaults or drawers in banks are, of course, the safest places to keep your valuables. The security is continuous, usually by armed guards and electronic surveillance devices and numerous locks and double locks. For your personal drawer only you and the bank attendants have keys and both keys are necessary to open the drawer. Very few bank vaults have been broken into in times of peace. However in times of war and complete chaos, the bank vaults are the prime places the enemy or looting mobs enter with plunder in mind. This, of course, is very rare but it has happened. An inconvenience of the bank vault is that it is only available during banking hours. Also, people observing the frequency of your goings and comings could possibly cause a risk. A distinct disadvantage of the bank vault is that the tax laws in many states make it mandatory for the bank officials to seal the bank drawer upon the death of the renter or anyone signed as co-renter. Usually the tax officials are notified and an inventory is taken so the proper inheritance tax can be assessed. In some instances, valuable papers and possessions are sealed in the drawer until the will can be probated. Also the rental cost per year, which can be from $20 to $50, should be a consideration.

The next most secure place, with the greatest amount of privacy, is a safe within your own home. A wall safe is the most convenient for you and also for a safe cracker. A safe, of any kind, is a great deterrent for any except the most professional burglar. Any safe or vault can be broken into. The great deterrent is to make it as difficult as possible for the burglar. The harder it is for a thief to gain access to valuables the more likely you are to keep them. Usually the thief goes for the easy marks. If many things of value are kept on the premises, an electronic or central alarm is advisable.

Much gold has been lost for lack of a good landmark.

Floor safes are made to be cemented into place in an existing floor. They are furnished with a concealing lid.

Courtesy Dick Hyatt, Kittle's Lock & Safe Co., Tucson, AZ.

Probably the safest of all, but possibly the least convenient to use, is the floor safe that is set in solid concrete flush with the surface of the floor. For many years commercial establishments have been using these for cash and valuables and the loss has been very small. This is due to several factors. First, the safe is sunken into the floor and cemented securely in place, usually making a smooth surface over which to lay a rug or slide a piece of furniture. A reliable safe and lock company can cut a piece of concrete out of an existing floor with a high speed abrasive saw and install the safe in minimal time and with minimum expense. A popular and convenient place to install one is in a clothes closet under a rug which can be folded back for easy access. A quick hit and run burglar seldom finds the safe there and if he does, he usually doesn't have the time, the tools or the knowledge to attempt to open it.

The lid is usually round and very thick. The safe is opened by lifting the lid out with a folding handle or bale. The entire lid, pins and lock assembly are of very hard steel that is almost impossible to drill. When the lid is closed it is seated on a hardened steel ring under which the two or three steel pins slide into a groove when locked. The round lid turns easily with the pins in the groove. Some safes have a removable combination knob which can be hidden elsewhere, thus frustrating a would-be burglar. Since the door fits snugly, and there is a very small crack around it, there is no place for a pry bar to be inserted; even if this were attempted, the fact that the lid turns easily makes it almost impossible to exert a pressure great enough to have any effect. Burglars attempt to open most floor and wall safes in this manner. An oxy-acetylene cutting torch could be used for access but the heat would destroy much of the valuable material inside. The drawbacks of the floor safe are the inconvenience of using it, the small door size and the usually small capacity inside the safe. The floor safe is your best bet for security and privacy especially if you don't mind getting down on your hands and knees and if you don't have much more than a cubic foot of gold to store.

The Equipment

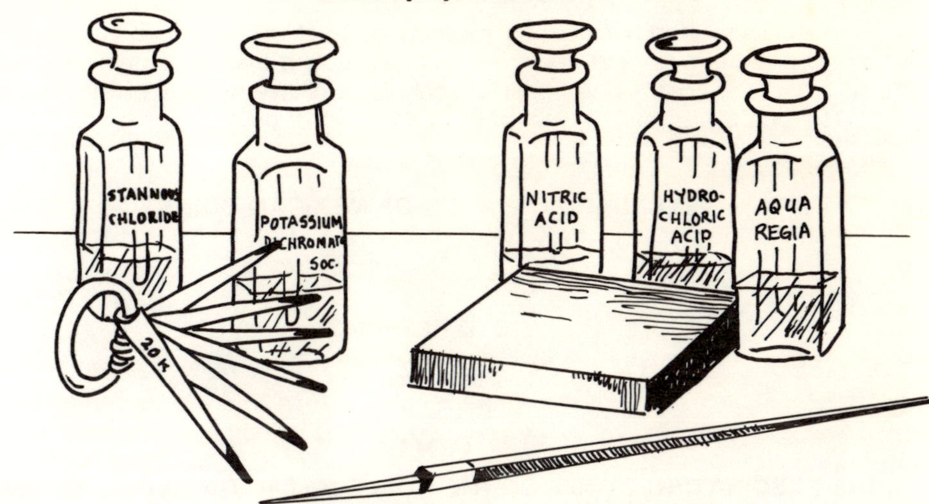

The chemicals, acids and equipment shown in this sketch are all that is necessary to perform the tests described in this book to identify gold, silver and platinum.

Acknowledgements

I would like to express my appreciation to the many friends who have freely given me encouragement, information and advice. I would like to show the greatest appreciation to these special people:

My greatest thanks to my wife Ethel Branson for helping me design, arrange and produce this book.

Connie Asch for her generous contribution of sketches, — her fine humor is a great addition to this book.

Sharleen M. Spivak for her help, advice and assistance.

Margaret Arozco,
G.I.A. Librarian
Paul A. Berquist
Mike Cuti
Don Green

Vern Davis
Ken Johnson
Reese Koonce
Nancie Mahan
Sterling Mahan

Earl Star
Shawn Star
Art Svoboda
Charles Vogt
William Wagner

The end of a good tale. Be careful with acids.

SUGGESTED READING

The books listed here are for persons seriously interested in obtaining more information about gold and silver. All of these books that were published before 1960 are out of print and will possibly only be available through public libraries.

The books are listed in the order of the date of publication.

STANDARD CATALOG OF MEXICAN COINS
by Dr. George W. Vogt 1978
Krause Publications
Iola, Wisconsin 256 pp.

GOLD REFINING
by George Gajda 1976
Santa Monica, CA
Technical — for the chemist 213 pp.

GOLD
THE FASCINATING STORY OF THE NOBLE METAL THROUGH THE AGES
by Daniel Cohen 1976
M. Evans and Company
New York, N.Y.

HOW TO BUY GOLD
by Timothy Green
Walker and Company 1975
New York, N.Y.
A very informative book for people who have money to invest in gold and what to expect from the investment.

EVERYTHING YOU WANTED TO KNOW ABOUT GOLD AND OTHER PRECIOUS METALS
by Russell Burkett 1975
Gem Guides
Whittier, CA 99 pp.
(Typewritten text, soft cover)

THE JEWELER'S MANUAL
by Richard T. Liddicoat, Jr. and Lawrence L. Copeland 1974
Gemological Institute of America
Los Angeles, CA 362 pp.

GOLDSMITHS AND SILVERSMITHS HANDBOOK
by George Gee
A Practical Manual for All Workers in Gold, Silver, Platinum and Paladium.
Second Edition Revised 1968
The Technical Press Ltd., London 109 pp.

GOLD COINS OF THE WORLD
by Robert Friedberg 1965
The Coin and Currency Institute
New York, N.Y. 415 pp.

GOLD RECOVERY — PROPERTIES AND APPLICATIONS
Edited by Edmund M. Wise 1964
D. Van Nostrand Co.
New York

JEWELERS WORKSHOP PRACTICES
by Leslie L. Linick
Henry Paulson & Co. 1948
Chicago
Trade secrets of Jewelry Manufacturers 516 pp.

TESTING PRECIOUS METALS
by C. M. Hoke
Gold, Silver, Platinum, Metals 1946
Identifying — Buying — Selling
A handbook for the jeweler, dentist, antiquarian, gold buyer, layman. 92 pp.

REFINING PRECIOUS METAL WASTES
by C. M. Hoke
Gold, Silver, Platinum, Metals 1940
A handbook for the jeweler, dentist, and small refiner. 362 pp.

THE METALLURGY OF GOLD
by Sir T. K. Rose and W. A. C. Newman
Seventh Ed.
J. B. Lippincot Co. 1937
Philadelphia 561 pp.

WORKING IN PRECIOUS METALS
by E. A. Smith
N.A.G. Press Ltd. 1934
London

A TEXTBOOK OF FIRE ASSAYING
by Edward E. Bugber
John Wiley & Sons 1922
New York 254 pp.

MANUAL OF ASSAYING
by Walter Lee Brown
Gold, Silver, Lead, Copper
12th Ed. E. H. Sargent & Co. 1907
London 590 pp.

THE METALLURGY OF GOLD
The metallurgical Treatment of Gold-bearing Ores
by M. Eissler
Crosby, Lochwood & Son 1889
London 340 pp.

METALLURGY — SILVER & GOLD
by John Percy
John Numary, LOndon 1880
The art of extracting metals from their ores. 700 pp.

A MANUAL OF PRACTICAL ASSAYING
by John Mitchell-Wm. Crookes
John Wiley & Sons 1873
New York

A TREATISE ON THE ASSAYING OF LEAD, COPPER, SILVER, GOLD NECESSARY
Translated by W. A. Goodyear
John Wiley & Sons 1868
New York